Bisesi and Kohn's

INDUSTRIAL HYGIENE EVALUATION METHODS

Second Edition

Bisesi and Kohn's

INDUSTRIAL HYGIENE EVALUATION METHODS

Second Edition

Michael S. Bisesi, Ph.D., CIH

Illustrated by Erik Bork

LEWIS PUBLISHERS

A CRC Press Company

Boca Raton London New York Washington, D.C.

Library of Congress Cataloging-in-Publication Data

Bisesi, Michael, S.
Bisesi and Kohn's industrial hygiene evaluation methods / Michael S. Bisesi.--2nd ed.
p. cm.
Previous ed. has title: Industrial hygiene evaluation methods.
Includes bibliographical references and index.
ISBN 1-56670-595-9 (alk. paper)
1. Industrial hygiene--Methodology. I. Bisesi, Michael S. Industrial hygiene evaluation methods. II. Kohn, James. P. III. Title.

RC967.B53 2003
616.9′803--dc21 2003051593

Direct all inquiries to CRC Press LLC, 2000 N.W. Corporate Blvd., Boca Raton, Florida 33431.

Lewis Publishers is an imprint of CRC Press LLC

International Standard Book Number 1-56670-595-9
Library of Congress Card Number 2003051593
Printed in the United States of America 1 2 3 4 5 6 7 8 9 0
Printed on acid-free paper

In Memory of

James P. Kohn

James P. Kohn, Ed.D., C.S.P., C.I.H., C.P.E. — friend, colleague and co-author of the first edition of this book — died in 1999. At that time, he was associate professor of Industrial Technology at East Carolina University (ECU). He earned his doctorate in Safety Management, with a minor in Experimental Psychology, from West Virginia University. Prior to joining the faculty at ECU, Dr. Kohn was an associate professor at Indiana State University, Terre Haute, and former corporate supervisor of health and safety training for a major Midwestern utility company. His specializations included: human factors/ergonomics, occupational health and safety training, industrial hygiene, and safety management theory and practice. He wrote several papers, textbooks and presentations in related areas. He was a certified safety professional (CSP), a certified industrial hygienist (CIH), and a certified professional ergonomist (CPE).

Dr. Kohn is survived by his wife Carrie and daughters Katy and Mary. He is deeply missed by family, friends, colleagues, and the occupational health and safety profession.

Preface

Evaluation of the occupational environment is one major focus of occupational health or industrial hygiene. Applicability of industrial hygiene theories, principles, and practices often extends beyond the occupational and manufacturing settings. Accordingly, many of the topics presented in this book are applicable to both occupational and nonoccupational and manufacturing and nonmanufacturing environments.

Numerous methods are used to evaluate the occupational environment and other applicable settings for physical, chemical, and biological agents and ergonomic factors that pose potential hazards and risks to the health of workers and others. This book is intended to summarize numerous industrial hygiene instruments and methods for evaluating occupational and some nonoccupational environments. It is divided into relatively short units and appendices that provide concise overviews and description of basic concepts, followed by technical exercises. Several units have example problems and some have short case studies.

The step-by-step exercises can be conducted in a setting where the applicable agents are detectable and measurable and other factors are observable. As an alternative, simulated evaluation exercises can be conducted in the classroom or laboratory. In either case, the learner will gain the knowledge, comprehension, and skill of basic industrial hygiene principles and practices for collection, detection, identification, calculation, and interpretation of qualitative and quantitative data.

The training exercises are intended only for demonstration purposes. Although based on and very representative of accepted industrial hygiene practices, the methods outlined here are not intended to be used as substitutes for those already established by various agencies and organizations. For two examples, learners are strongly encouraged to be familiar with practical information and methods summarized in publications such as the *NIOSH Manual of Analytical Methods* and the *OSHA Technical Manual*. In addition, for more expansive coverage of the related theoretical aspects, learners using this book should be or become familiar with several of the excellent recommended references listed in the bibliography. This book is an excellent adjunct to books addressing comprehensive practice, such as *The Occupational Environment — Its Evaluation and Control* and the *Fundamentals of Industrial Hygiene*. Finally, training exercises should be conducted under the supervision of an experienced industrial hygienist while ensuring that personal health and safety precautions are implemented.

This book is an introductory, condensed learning reference intended for novice environmental and occupational health and safety undergraduate and graduate students and practitioners. It also will serve as a useful book for the nonacademic continuing education and professional development of individuals who need an understanding or review of evaluation methods fundamental to industrial hygiene practice. The first edition was well received as a succinct review for even experienced practitioners. This second edition is expanded and improved and will certainly serve and accomplish the same, if not more.

Michael S. Bisesi, Ph.D., C.I.H.

Acknowledgments

I extend my sincere appreciation to my wife Christine and my sons Antonio (Nino) and Nicolas (Nico) for their unconditional love and support in all that I do. I also recognize my parents Anthony (deceased) and Maria Concetta Bisesi for the opportunities they once provided for and the work ethic they instilled in me.

In addition, I acknowledge the following mentors: Rev. Francis Young, Dr. George Berkowitz, Dr. Raymond Manganelli, Dr. Barry Schlegel, Dr. John Hochstrasser, Dr. Richard Spear, Dr. Herman Koren, Dr. Roy Hartenstein, Dr. Christopher Bork, and Dr. Keith Schlender for sharing their knowledge, wisdom and encouragement during various phases of my academic and professional journeys.

Also, I want to thank the professionals who reviewed units within this book: my wife Christine Bisesi, M.S., C.I.H., C.H.M.M.; colleagues Farhang Akbar, Ph.D., C.S.P., C.I.H., Manny Ehrlich, M.S., M.Ed., Brian Harrington, Ph.D., M.P.H., Sheryl Milz, Ph.D., C.I.H., and Sandra Woolley, Ph.D.; and graduate students Susan Arnold, C.I.H. and Cristine Amurao, M.D., M.P.H. Finally, I thank Erik Bork for his computer illustrations of the various instruments depicted in this book.

About the Author

Michael Bisesi, Ph.D., C.I.H., is an environmental and occupational health scientist and board certified industrial hygienist working full-time as professor and chairman of the Department of Public Health in the School of Allied Health at the Medical College of Ohio (MCO). He has a joint appointment in the Department of Pharmacology in the School of Medicine, serves as the associate dean of Allied Health Programs, and is a director of the Northwest Ohio Consortium for Public Health. At MCO, he is responsible for research, teaching, service, and administration. He teaches a variety of graduate-level and continuing education courses, including toxicology, environmental health, monitoring and analytical methods, and hazardous materials. His major laboratory and field interests are environmental toxicology involving biotic and abiotic transformation of organics, fate of pathogenic agents in various matrices, and industrial hygiene evaluation of airborne biological and chemical agents relative to human exposure assessment. He periodically serves as a consultant via Enviro-Health, Inc., Holland, Ohio.

Dr. Bisesi earned B.S. and M.S. degrees in environmental science from Rutgers University and a Ph.D. degree in environmental science from the SUNY at Syracuse in association with Syracuse University. He continues to complete additional post-graduate coursework in epidemiology and molecular biology. He also completed a fellowship at MCO in Teaching and Learning Health and Medical Sciences and earned a graduate certificate.

Dr. Bisesi has published several scientific articles and chapters, including two chapters in the *Occupational Environment: Its Evaluation and Control* and three chapters in 4th and 5th editions of *Patty's Industrial Hygiene and Toxicology.* In addition, he is the author of the textbook *Industrial Hygiene Evaluation Methods* and co-author of two other textbooks, the *Handbook of Environmental Health and Safety, Volume I: Biological, Chemical and Physical Agents of Environmentally Related Disease* and *Volume II: Pollutant Interactions in Air, Water and Soil.* He is a member of the American Public Health Association, American Industrial Hygiene Association (Fellow), American Conference of Governmental Industrial Hygienists, National Environmental Health Association, and the Society of Environmental Toxicology and Chemistry.

Contents

UNIT 1

Evaluation of Hazardous Agents and Factors in Occupational and Nonoccupational Environments

LEARNING OBJECTIVES

At the completion of Unit 1, including sufficient reading and studying of this and related reference material, learners will be able to correctly:

- Summarize the roles of occupational health specialists/industrial hygienists.
- Name and define the categories of industrial hygiene sampling and analysis for determination of external exposure to physical, chemical, and biological agents.
- Discuss the aspect of evaluating ergonomic factors.
- Discuss the aspect of evaluating hazard controls.
- Discuss the industrial hygiene and medical monitoring and analysis relative to external, internal, and effective exposures.

OVERVIEW

The occupational environment can be simply defined as any place, indoors or outdoors, where people work in return for financial or other remuneration. The profession of occupational health, or industrial hygiene, is based on the tetrad of anticipation, recognition, evaluation, and control of agents, factors, and stressors related to the occupational environment that may adversely affect the health of workers and other members of the community. All four aspects of the tetrad are interrelated. The content of this book, however, will mostly emphasize some major instrumentation, methods, and practices of industrial hygiene evaluation.

The occupational environment is evaluated by occupational health specialists, historically and presently, most commonly referred to as industrial hygienists. Industrial hygienists are involved with evaluation of the occupational environment from several different perspectives. Industrial hygiene evaluation activities include conducting walk-through surveys of facilities and applicable monitoring activities to gather both qualitative and quantitative data. Monitoring activities include sampling and analysis to collect, detect, identify, and measure hazardous physical, chemical, and biological agents present in specific areas and to which workers and others are potentially or actually exposed. Industrial hygienists commonly interact directly or indirectly with clinical professionals, such as physicians, nurses, and audiologists. Industrial hygienists are often familiar with some general principles of clinical techniques to evaluate workers to determine if they reveal signs of adverse impact from excessive exposure to agents, such as hearing loss due to prolonged exposure to elevated sound levels. In addition, industrial hygienists evaluate control measures, including work practices, personal protective equipment, and ventilation systems, to determine if they effectively reduce the potential for worker exposure. Industrial hygienists also may be involved with evaluation of ergonomic factors to determine if there is an appropriate match or fit between workers and their physical workplace environments.

Industrial hygiene principles and practices often extend beyond the occupational and manufacturing settings. For example, professional- and technician-level industrial hygienists are frequently involved in indoor air quality (IAQ) investigations in nonoccupational settings, such as homes, schools, and other nonmanufacturing settings. Accordingly, many of the topics presented in this book are applicable to both occupational and nonoccupational and manufacturing and nonmanufacturing environments.

EVALUATION OF EXTERNAL EXPOSURE VIA INDUSTRIAL HYGIENE MONITORING

Sampling and analysis refer to the representative collection, detection, identification, and measurement of agents found in environmental matrices such as air, water, and soil. In occupational and nonoccupational environments, both indoors and outdoors, air is sampled (collected) to detect and identify physical, chemical, and biological agents and to measure related levels. The most common matrix that is sampled and analyzed in the occupational environment is the air. Indeed, inhalation of contaminated air by workers is considered the major mode of foreign agent entry in most occupational environments. In addition, the air serves as a matrix for elevated sound levels, extremes of temperature and humidity, and transfer of ionizing and nonionizing radiation energies.

The data collected and analyzed are used to evaluate both actual and potential external exposures to agents encountered by humans. In turn, the levels are compared to established occupational exposure limits (OELs) to determine if acceptable values for exposure have been exceeded. In the U.S., regulatory permissible exposure limits (PELs) are enforced by the Occupational Safety and Health Administration (OSHA) for nonmining operations and processes. Regulatory exposure limits for the mining industry are the threshold limit values (TLVs) enforced by the Mine Safety and Health Administration (MSHA). Other agencies have established occupational exposure limits as guidelines, most notably the TLVs by the American Conference of Governmental Industrial Hygienists (ACGIH) and Recommended Exposure Levels (RELs) by the National Institute for Occupational Safety and Health (NIOSH). In some cases, a parameter will be monitored without concern necessarily for excessive exposure. For example, illumination is evaluated and compared to recommend guidelines to assure that there is an appropriate, neither inadequate nor excessive, quantity of lighting. Refer to Appendix A for a summary of some related strategies for exposure assessment and calculations of time-weighted averages (TWAs) for comparisons to established regulatory and nonregulatory occupational exposure limits. In relation, Appendix B is an example of an outline format of an industrial hygiene evaluation report showing information that must be considered, documented, and reported.

Sampling and related analytical activities are divided into several categories to reflect the type of monitoring that is conducted. Categories are based on factors that include time, location, and methods of collection (sampling) and detection, identification, and measurement (analysis). Each serves a purpose in evaluating the occupational environment to determine the degree of workers' external exposure to various agents.

(i) Instantaneous or Real-Time Sampling

Instantaneous sampling refers to the collection of a sample for a relatively short period ranging from seconds to typically less than 10 min. A major advantage of instantaneous sampling is that both sample collection and analysis are provided immediately via direct readout from the sampling device. The data represent the level of an agent at the specific time of sampling. Accordingly, instantaneous sampling is also referred to as direct reading and real-time sampling. Real-time sampling is perhaps a more appropriate designation since there are some devices already developed and being designed for integrated or continuous monitoring (see Section ii) that provide a direct-readout or instantaneous result without need for laboratory analysis and the associated delays. In addition, the main purpose of real-time sampling is to reveal what a level of an agent is, at an immediate point of time or during real-time.

The application of real-time sampling varies. The strategy is used when preliminary information regarding the level of an agent is needed at a specific time and location. For example, real-time sampling is commonly used for screening to identify agents and measure related levels. This is important for developing follow-up monitoring strategy and determining if integrated sampling is warranted. Real-time sampling is also beneficial for determining levels of agents during short-term operations or specific isolated processes when peak levels are anticipated or suspected.

(ii) Integrated or Continuous Sampling

Integrated sampling refers to the collection of a sample continuously over a prolonged period ranging from more than 10 or 15 min to typically several hours. Integrated sampling is also referred to as continuous monitoring reflective of the extended period of sample collection. Most work shifts are 8 h and occupational exposure limits are most commonly based on an 8-h exposure period. Accordingly, it is very common as well for sampling to cover the duration of the shift. Several strategies can be followed. For example, a sample run could be started immediately at the beginning of the 8-h shift and allowed to run until the shift ends. Analysis of the sample would provide a single value representative of the level of a particular agent during the shift. The single value represents an integration of all the levels during the shift. The single value, however, does not provide information regarding fluctuations of levels that were higher, lower, or not detectable during shorter periods within the 8-h shift. In addition, there is no indication as to the levels at specific times and locations during the shift. As a result, an alternative strategy could involve collection of several samples of shorter duration during the entire shift. In turn, analysis of the individual samples provides levels associated with specific times, tasks, and locations during the 8-hour shift. Concentration (C) and corresponding sample time (T) data from one sample (C_1) or several individual samples (C_1 to C_n) can be time-weighted ($C \times T$) and averaged, by dividing by a specific time period (e.g., 8 h), to provide a single overall TWA for the 8-h shift (Appendix A).

$$\begin{aligned} 8\text{ h TWA} &= \frac{(C_1T_1)}{8\text{ h}} \\ \text{or,}\quad 8\text{ h TWA} &= \frac{(C_1T_1)+(C_2T_2)+\cdots+(C_nT_n)}{8\text{ h}} \end{aligned} \tag{1.1}$$

A major advantage of integrated sampling is that it provides a single value for the level of an agent over a prolonged period. The level of an agent can be determined during discrete times and locations within a workshift to assist in identifying factors that influence elevated values of exposure or external exposure. A major disadvantage associated with integrated sampling is that in most cases, samples must be submitted to a laboratory for analysis prior to knowing what has been detected in the related measurements. This frequently results in a delay between sample collection and data reporting.

Several fundamental procedures must be followed when conducting integrated personal and area sampling. It is important to assure that monitoring devices and/or sampling trains are properly assembled, calibrated, and operated for the specific monitoring activity. It is equally important to assure that field monitoring data are recorded so that samples can be associated with specific locations, areas, individuals, dates, times, processes, equipment, temperatures, humidity, atmospheric pressure, and so on. Figure 1.1 summarizes a representative example of a generic integrated monitoring protocol.

- ☐ Make sure that all active monitoring devices are pre-calibrated and order checked for accurate calibration. Check batteries for charge. In addition, confirm that sampling media are not expired.
- ☐ Select a worker or area to be monitored. Briefly explain the purpose of monitoring to the worker and/or workers in the area. Advise individuals not to tamper with the instrument or medium. Record the worker's name or the area sampled, worker's social security number, and job title. Record date, sampling location and sampling device and/or media identification number.
- ☐ If a personal sample, attach the monitoring device and/or medium to the worker and make sure that they do not interfere with the worker's activities. Attach the sampling medium (e.g., filter cassette) at the worker's clavicle near the collar if a breathing zone sample is needed. If a hearing zone sample is needed, attach the medium (e.g., audiodosimeter microphone) at the trapezius or the ear. If a flexible hose or a cord is involved, allow enough slack to accommodate worker's range of motion when standing, sitting, bending, and twisting and secure any excess so that it does not serve as a potential interference or hazard.
- ☐ For an area sample, position the monitoring device and/or medium approximately 4 to 6 feet from the floor. Make sure that the medium is not in direct contact with or too close to a contaminant (e.g. settled or spilled particulate)
- ☐ Turn "ON" an active monitoring device and record "start time". Make sure that the device is operating. If a passive device is used, record "start time" when it is first exposed to the workplace atmosphere.
- ☐ Document the worker's performed tasks and/or processes operating in the area during the monitoring period. Note times when exposure may be high due to specific activities or process phases. Check the monitoring device and medium after the first 15 minutes to half an hour, and at least two-hour intervals thereafter. If applicable, change the medium when conditions warrant (e.g., signs of overloading; excessive airborne concentrations of contaminant; to isolate exposure to specific time periods or specific tasks).
- ☐ Record the "stop time" when medium is changed (or sampling is concluded) and "start time" when medium is replaced; make sure that the identification number of each sample is recorded. If applicable, handle field blanks in a similar manner as samples (recording start and stop times in military time makes it easier to determine sample time and convert to minutes).
- ☐ Remove monitoring device and/or media from the worker or the area being sampled.
- ☐ Make sure that all active monitoring devices are post-calibrated or checked for accurate calibration. Post-calibration should be conducted prior to recharging batteries.

Figure 1.1 Example of a protocol for personal and area integrated air sampling.

(iii) Personal Sampling

Personal sampling involves direct connection of an integrated monitoring device to a worker. The device, in turn, will collect a sample or record the intensity of an agent in the specific areas and during specific tasks conducted by a worker. Indeed, personal sampling is frequently a form of mobile monitoring since the sampling device travels to the same areas and at the same times as the worker that wears it.

If inhalation is the mode and the respiratory system the route of entry of an agent, the sampling device or related sampling medium is positioned in the worker's breathing zone. The breathing zone refers to an area within a 9- to 12-in. distance (radius) from the worker's nose and mouth (Figure 1.2a). Typically, an integrated monitoring device for personal sampling is attached near the worker's clavicle (collar bone). Relative to evaluation and impact of sound levels, however, hearing is the major mode and the auditory system is the route of entry. When conducting personal monitoring for sound, therefore, the sampling device can be connected to the worker's hearing zone. This zone is ideally the ear itself, or the trapezius region of the shoulder — a region within a 9- to 12-in. distance from the ear (Figure 1.2b).

Instantaneous or real-time monitors also can be used to determine levels of agents in a worker's breathing and hearing zones. For example, a sound level meter can be held by the individual

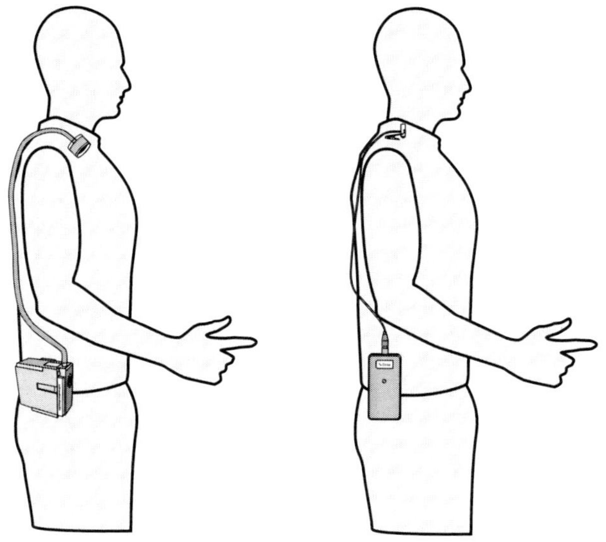

Figure 1.2 (a) Active-flow pump is positioned at the worker's waist. Flexible tubing connects the pump to a sample collection medium clipped to the worker's shirt along the collar bone and positioned within his breathing zone to collect a personal integrated sample. (b) Sound (noise) dosimeter is positioned at the worker's waist. Flexible wire connects the dosimeter to a microphone clipped to the worker's shirt along the trapezius and positioned within his hearing zone to collect a personal integrated sample.

conducting the monitoring in the auditory region of a worker. This would provide an instantaneous assessment of sound levels in the worker's hearing zone at the specific time of monitoring.

(iv) Area Sampling

The focus of area sampling is to evaluate the levels of agents in a specific location, instead of evaluating levels encountered by a specific worker. Area integrated monitoring devices are typically positioned in a stationary location (Figure 1.3a). Stationary area integrated samples are often collected at a height of approximately 4 ft from the floor or ground. The data from a stationary area integrated sample represent the level of an agent in the specific area during the sampling period. Area instantaneous or real-time monitoring, however, involves area sampling in either a stationary or mobile mode. Stationary area instantaneous monitoring may be conducted while positioning or holding and operating a direct read instrument and standing still in a given location. Alternatively, an instantaneous monitoring instrument can be transported via carrying it or rolling it on a cart to various locations while intermittently checking the indicator or readout on the instrument (Figure 1.3b).

(v) Active Flow Sampling

Presently, most monitoring techniques for actual collection of an air sample or contaminant from the air involve active flow methods. Active flow sampling implies that energy, such as an electronically powered (either AC or DC) device, is required to collect the sample. Air and airborne contaminants are actively pulled through a collection medium or into a collection container. For example, battery-powered air sampling pumps are frequently used to pull air through sample collection media or into a sample container. Energy also can be generated manually, via physically pumping a device for example, to conduct active flow sampling.

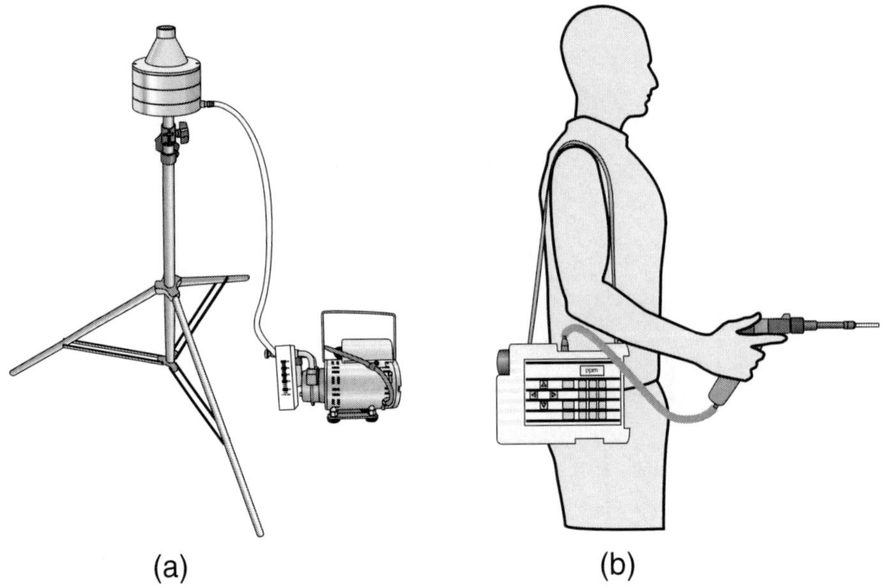

Figure 1.3 (a) Two-stage bioaerosol impactor is positioned on a stationary tripod. Flexible tubing connects the active-flow pump to the impactor to collect an area integrated sample. (b) Active-flow, direct read-out instrument is carried and operated by an individual walking from area to area within a facility to conduct mobile area sampling. Flexible tubing connects the instrument to a probe pointed in the air to collect area instantaneous samples.

(vi) Passive Flow Sampling

Passive flow sampling implies that neither electrical nor manual energy is required to operate the air sampling device. The method applies to the collection of diffusible gases and vapors; collection of settling particulates; measurement of temperature, pressure, and humidity; and the detection and measurement of forms of ionizing and nonionizing radiation. In the case of gases and vapors, collection using a passive monitor or dosimeter (if used for personal monitoring) relies on the movement (diffusion) of a gas or vapor from an area of relative high concentration, such as the air, to an area of relatively low concentration, the passive monitor. There is no need to actively move the air so that it and the contaminants flow into or through a sample collection medium. As for other examples, although not commonly used for evaluation of the occupational environment due to poor accuracy and precision, settleable dust can be collected passively using a pre-weighed dust fall jar or settleable microorganisms collected passively in open culture dishes filled with agar growth media.

(vii) Surface Sampling

Surfaces that are potentially or suspected to be contaminated with a toxic or pathogenic agent are sampled. Moistened or pre-treated cellulose (paper) sheets (wipes) and sponge-cotton-tipped swabs are commonly used to collect a sample from a surface. The media are then analyzed for the contaminant of interest. Surface sampling is also referred to as wipe or swab sampling.

(viii) Bulk Sampling

Bulk sampling refers to collection of a representative portion of a matrix. For example, there are times when actual collection of the air, not simply the contaminant separated from the air, is warranted. Accordingly, a special glass or metal cylinder or plastic bag (e.g., Mylar) may be used to collect a bulk sample of air. Aliquots (subsamples) of the bulk sample can be analyzed for specific agents. Bulk sampling also refers to the collection of a real or potential source of an agent. For example, bulk samples of building materials are commonly collected and analyzed for asbestos fibers. In addition, bulk samples of liquid solvents or powdered materials may be collected for analysis to determine the identity and percentage of the toxic ingredients. Bulk samples of powdered mineral materials are commonly collected and analyzed for crystalline silica.

(ix) Grab Sampling

Grab sampling refers to collection of a sample at a specific location and specific time. Thus, real-time sampling is a form of grab sampling the air. The method, however, also can pertain to the random collection of samples without regard to a specific time or location.

EVALUATION OF ERGONOMIC FACTORS

The relationship between workers and their workplace surroundings is another important aspect that must be evaluated. Ergonomic evaluations involve measurements of physical aspects of the workers and the tools, machinery, and workstations they use. Ideally, the fit between workers and the instruments, equipment, and settings of their occupations should result in minimal discomfort to them and increased efficiency and effectiveness.

EVALUATION OF CONTROLS

Control of human exposure to hazardous agents and factors is an important preventative and remediative aspect in occupational and nonoccupational environments. Two major measures to control external exposures to contaminants, especially airborne, are use of personal protective equipment and local and general ventilation systems. Inherent to virtually every hazard control program, however, is also the expectation for individuals, both workers and nonworkers, to implement and follow appropriate, feasible, and reasonable practices to minimize or even eliminate their personal exposures to various hazardous agents and factors.

Industrial hygienists must evaluate occupational settings where personal protective equipment is used, including materials that serve as barriers between contaminants and the ocular (eyes), dermal (skin), and respiratory systems, to determine if there is proper selection, maintenance, fit, and use. A deficiency in any of these areas can certainly compromise the protection that should be provided by the equipment.

Local and general ventilation systems are the primary engineering controls to reduce or eliminate airborne contaminants and to provide atmospheric comfort in occupational and nonoccupational environments. Industrial hygienists should be directly or at least indirectly involved with periodic evaluation of ventilation systems to assure that the design and operation are efficient relative to the characteristics of the environmental setting, including the number of workers and occupants and the types and levels of contaminants present.

EVALUATION OF INTERNAL OR EFFECTIVE EXPOSURE VIA MEDICAL MONITORING

Ideally, industrial hygiene evaluation data are used proactively to prevent occupationally related diseases or dysfunctions. It is hoped that measurement of agents and observation of factors that are shown to be at hazardous levels or conditions can be remediated and controlled prior to the initiation of any adverse impact to workers. The data, however, mostly represent agents or factors to which workers and others are externally exposed. External exposure refers to the initial contact between the human body and a contaminant in an environment. Industrial hygiene practices contribute to the characterization of external exposures. In turn, medical evaluation of workers is warranted to determine if workers are medically qualified for certain tasks and if they exhibit signs of occupationally related diseases or dysfunctions due to excessive external exposures (i.e., causes) and potentially corresponding internal and effective exposures (i.e., effects). Internal exposure suggests that an agent has absorbed or penetrated into the body following external exposure. Effective exposure, in turn, refers to signs or symptoms of adverse biochemical, physiological, morphological, and psychological impact resulting from excessive external and related internal exposures and interactions.

Many of the existing OSHA standards in the U.S. mandate various medical surveillance activities, including, but not limited to, physical examinations, x-rays, biological sampling and analysis of blood and urine, pulmonary function testing of the respiratory system, and audiometric testing of the auditory system. The combination of industrial hygiene and medical monitoring activities and interactions provides comprehensive data, based on both workers' and others' external and internal exposures to various agents encountered in occupational and nonoccupational environments.

UNIT 2

Hazardous Environmental Agents and Factors

LEARNING OBJECTIVES

At the completion of Unit 2, including sufficient reading and studying of this and related reference material, learners will be able to correctly:

- Categorize and summarize the general characteristics of major physical agents and the potential impact to human health.
- Categorize and summarize the general characteristics of chemical agents and the potential impact to human health.
- Categorize and summarize the general characteristics of biological agents and the potential impact to human health.
- Categorize and summarize the general characteristics of ergonomic factors and the potential impact to human health.

OVERVIEW

Numerous agents and factors are present in occupational and nonoccupational environments that can cause adverse health effects to workers and others who are exposed to elevated levels and unacceptable conditions. In addition to general guidelines, specific occupational exposure limits have been established for many of the hazardous agents to limit the degree and duration of worker exposure. Monitoring is conducted, especially relative to the air matrix, to determine the levels of various physical, chemical, and biological agents present in the workplace and elsewhere. Guidelines also have been established relative to factors other than exposure to physical, chemical, and biological agents that can cause adverse impact to human health. Industrial hygiene professionals must have at least the fundamental knowledge and comprehension of major categories and types of hazardous agents and their relationships to human diseases and dysfunctions. Accordingly, four major categories of environmental agents and factors of concern (i.e., physical, chemical, biological, and ergonomic) are succinctly summarized in Table 2.1.

PHYSICAL AGENTS

Physical agents include energy sources such as sound, nonionizing radiation, ionizing radiation, and temperature extremes. The hazards associated with these agents are typically associated with the exposure to and impact of a form of airborne physical energy on the human system.

Table 2.1 Categories of Hazardous Environmental Agents and Factors

Physical Agents
Sound
Nonionizing radiation
Ionizing radiation
Temperature extremes (e.g., heat stress)
Chemical Agents
Toxic
Flammable
Corrosive
Reactive
Biological (Microbial) Agents
Pathogenic
Infective
Allergenic
Intoxicating
Ergonomic Factors
Workplaces
Layout and design
Workers
Anthropometrics

(i) Sound

Sound waves are disturbances in the uniform pressure of air molecules, detectable by the human ear, caused by vibrating objects. As objects vibrate back and forth, they alternately cause compressions and rarefactions (partial vacuums) of molecules in the air. Thus, sound is an energy form of vibration that may be conducted through various media including solids, liquids, or gases. The velocity of sound wave travel varies depending upon the mass (density of the medium) and elastic reactions (pressure) of the molecules.

The amplitude or intensity of sound is related to sound pressure. Sound pressure is the deviation of air pressure from normal atmospheric pressure and is related to the amplitude of sound. The best value to use to determine sound pressure amplifications would be average pressure changes. However, one problem would be created if average pressure were used; average compression and rarefaction pressure change always equal zero or atmospheric pressure. Accordingly, in place of simple average, instantaneous pressures are first squared and then converted to a value referred to as the root mean-square (RMS). RMS sound pressures are measured in dynes per square centimeter (dyn/cm^2), newtons per square meter (N/m^2), microbars (μbar), or more commonly micropascals (μPa). Sound power is another measure associated with the amplitude of a sound wave. It refers to the total sound energy radiated by the source. Sound power is typically expressed in the unit watts.

Sound wave amplitude may be defined either as the quantity of sound produced at a given location away from the source or in terms of its overall ability to emit sound. It is the amount of pressure change from the average atmospheric pressure, which is generally described by sound pressure or sound intensity. The loudness of sound is directly related to the intensity of sound. Subjectively, loudness is a term that refers to an observer's perception or impression of loud vs. soft (quiet) sounds.

Associated with the measurement of sound pressure and sound power is the logarithmic unit decibel (dB). By definition, the decibel is a dimensionless logarithmic unit relating a measured quantity such as sound pressure in micropascals or sound power in watts (W) to a reference quantity. Decibels are commonly used to describe levels of acoustic intensity, acoustic power, and hearing thresholds. Sound pressure levels are measured in decibels referenced to 20 μPa. The reference value corresponds to the lowest sound pressure that will produce an audible sound that is detectable by humans.

$$\begin{aligned} SPL &= 20 \log \frac{P_{meas}}{P_{ref}} \\ &= 20 \log \frac{P_{meas}}{20\ \mu Pa} \end{aligned} \tag{2.1}$$

In addition to amplitude or intensity of sound waves, three other important characteristics of sound waves that are useful in understanding and evaluating sound in the occupational environment are frequency, wavelength, and speed (velocity). Frequency of sound waves is an important attribute that must be determined. Frequency (f) is the number of complete vibrations within a given period of time. It is the total number of completed cycles (compressions and rarefactions) of a sound wave per second (cps) and is measured in units of hertz (Hz). Higher frequencies (commonly described as high-pitched sounds) tend to be more hazardous than low-pitched sounds.

Another sound wave attribute of importance is wavelength. Wavelength refers to the distance required for one complete pressure cycle to be completed. This attribute is important when considering the effects of sound waves. For example, sound waves that are larger than surrounding objects will diffract or bend around these obstacles. Small sound waves, on the other hand, will refract or be dispersed, and surrounding obstacles can function as effective sound barriers. Wavelength is an important variable in noise control activities.

The final parameter of sound that must be considered is the speed of sound or sound wave velocity. Speed is dependent upon the ambient temperature of the medium and its environment, and at about 22°C the speed of a sound wave is approximately 344 m/sec. The speed (c) of a sound wave is determined by the product of its wavelength (λ) and frequency.

$$\begin{aligned} c &= \lambda f \\ 344\ \mathrm{m/s} &= \lambda f \end{aligned} \tag{2.2}$$

Excessive or unwanted sound is referred to as noise. Sound energy travels as waves through air and is generated as a result of the oscillation of sound pressure above and below atmospheric pressure. The sound energy interacts with the human system by initially vibrating the tympanic membrane or eardrum, which conducts the energy to the middle ear where the energy is conducted further via vibration of the ossicles. Ultimately, the energy enters the inner ear where pressure changes in the cochlea and movement of hair cells present cause stimulation of the auditory nerve which transmits the signal to the brain. Excessive levels of sound at a given range of frequencies, combined with a related excess in duration of exposure, can result in damage to the auditory system via conductive or sensorineural hearing loss. Units 16 and 17 in this book summarize theories, instrumentation, and methods of monitoring applicable to sound. Unit 18 summarizes the theory, instrumentation, and method to evaluate sound (noise)-induced hearing loss.

(ii) Nonionizing Radiation

Radiation is often described as energy in motion, either in the form of electromagnetic waves or very small, high-speed particles. Electromagnetic waves are characterized by three related factors: frequency, wavelength, and speed. Frequency is the number of complete wave cycles per unit period of time, often in cycles per second knowns as hertz. Wavelength is the length of one complete wave cycle, often expressed in meters (m), millimeters (mm), and kilometers (km). Speed is the time rate change of displacement at which a wave travels. Speed is constant for a given media, and in the vacuum (free-space), it is approximately 3×10^8 m/sec. The three factors of frequency, wavelength, and speed are related; the speed is equal to the frequency times the wavelength.

$$c = \lambda f$$
$$3 \times 10^8 \text{ m/s} = \lambda f \tag{2.3}$$

When moving between different types of media, such as solids, liquids, gases, and vacuum, all electromagnetic waves exhibit certain physical phenomena including reflection, refraction, diffraction, absorption, and transmission. Electromagnetic waves consist of oscillating electric and magnetic fields (in the form of vectors) that are perpendicular to each other and each at a right angle (90°) to the velocity and direction of the travelling wave. Velocity is vector form of the speed. Electromagnetic waves in certain wavelengths make up a spectrum of waves or spectrum categories.

The spectrum categories of electromagnetic waves differ based on ranges of wavelengths and, therefore, frequencies. In addition, the categories of the electromagnetic spectrum may be differentiated by the quantity of photon energy. Photon energy is measured in units of joules (J), which refer to the work associated with a 1-N force moving an object a distance of 1 m. The photon energy (E) is related to the product of Planck's constant (h; 6.624×10^{-34} J/sec) and the frequency (cycles per second; cps) of electromagnetic wave.

$$E = hf$$
$$= (6.624 \times 10^{-34} \text{ J/sec}) \text{ (cps)} \tag{2.4}$$

Electromagnetic waves also behave similar to discrete packets of energy termed photons. The photon energy of electromagnetic radiation can be expressed in units of electron volts (eV). One eV equals the amount of energy gained by an electron when passing through an electric potential difference of 1 V. In terms of conventional energy units, 1 eV is equivalent to 1.602×10^{-19} J. The power of electromagnetic radiation is related to energy over time and is measured in units of watts. One watt represents the total energy (J) over the unit of time (sec) and is equivalent to 1 J/sec.

In the field of health and safety, the electromagnetic waves are usually divided into two major spectrum categories: ionizing radiation and nonionizing radiation. The primary difference between ionizing radiation and nonionizing radiation is that ionizing radiation has sufficient energy (>10 eV) to dislodge subatomic orbital electrons upon impact, thus causing ions to form. Nonionizing radiation has lower energies and is therefore incapable of dislodging orbital electrons. Nonionizing radiation can become harmful by depositing thermal and vibrational energies within materials and leaving atoms in an excited state. Nonionizing (electromagnetic) radiation, in order of increased wavelengths (decreased frequencies and therefore decreased photon energy), is divided into spectrum subcategories of ultraviolet (UV), visible light, infrared (IR), radiofrequencies (RF), and extremely low frequency (ELF) electromagnetic fields (EMF).

The biological effects associated with exposure to nonionizing radiation vary depending upon the specific spectrum category of concern and the organs of the body exposed. The organs of greatest concern associated with exposure to visible light are the eyes and surrounding tissues. Acute effects include retinal burns, eyestrain, tearing, blurring, and headaches. Chronic effects of exposure to high intensities of visible light include degeneration of the cones of the retina. The organs of greatest concern associated with exposure to UV and IR are the eyes and skin. Skin erythema and burning as well as lens and corneal (including photokeratitis and photoconjunctivitis) effects have resulted from exposure to specific wavelengths. Cataracts and welder's flash are possible eye injuries resulting from UV and IR radiation.

The spectrum region of RF is characterized by relatively short frequencies, and covers frequencies from 300 GHz to 3 kHz. Microwave radiation (MW) is a subset of RF, but usually this range of frequencies is considered to be separate spectral regions: MW from 300 MHz to 300 GHz (1 mm to 1 m) and radiowaves from 3 kHz to 300 MHz (1 m to 100 km). MW and RF can penetrate the human body and cause adverse effects to internal organs such as thermal strain, teratogenesis, and ocular effects (cataracts and cornea damage). There is increasing evidence that cellular reactions to MW and RF may also occur causing immunological, central nervous system, neuroendocrine, and arthritis disorders.

ELF radiation has elicited increased concern, especially frequencies of 60 Hz. Sources of ELF include high voltage power lines, video display terminals (VDTs), and close proximity to electrical wiring such as electric appliances. Cancer, such as leukemia, has been suggested as one of the potential biological effects according to preliminary research of ELF exposure. This issue remains open for debate, and external exposures to ELF radiation and the impact to health remain in need of further investigation.

Units 20, 21, and 22 summarize the theories, instrumentation, and methods of monitoring applicable to some types of nonionizing radiation.

(iii) Ionizing Radiation

Atoms with unstable nuclei will decay due to the imbalance of protons (p^+) and neutrons (n^o) and will emanate energy in the form of ionizing radiation. These nuclides undergo transformations to achieve stability via emanation of atomic energy in the form of particulates and electromagnetic photons. The particles and photons impart energy to matter with which they interact, resulting in ionization.

Ionizing particulate radiation consists of alpha particles (2 n^o + 2 p^+) in the form of a charged helium nucleus (${}_{2}^{4}He^{+2}$), beta particles (negatron as e^- and positron as e^+), protons, and neutron particles. Ionizing electromagnetic wave radiation consists of photon energies as x-rays and gamma rays. The particulate and electromagnetic wave forms of ionizing radiation can interact via direct or indirect ionization of macromolecular or cellular components of the human body which, like those materials classified as toxic, may result in adverse biochemical and physiological changes manifested as abnormalities, illnesses, or premature deaths among those exposed or their offspring.

The energy of a photon or particle is measured in electron volts. The rate at which an element undergoes radioactive decay is referred to as activity and is measured in units of curies (Ci) or becquerel (Bq). One curie is equivalent to 3.7×10^{10} disintegrations per second (dps). The unit becquerel (Bq) is a contemporary alternative to curies, where 1 Bq = 1 dps and, therefore, 1 Bq = 2.7×10^{-11} Ci. Half-life refers to the time required for 50% of a radioactive element to decay (transform into its daughter nuclei) and, accordingly, the activity to decrease.

Exposure, relative to ionizing radiation, is the intensity of an x-ray or gamma ray photon beam determined by the amount of ionization (electrical charge) per mass or volume of air, measured in units of roentgens (or coulombs [C] per kilogram of air, where 1 R = 2.58×10^{-4} C/kg of air). It takes approximately 34 eV of energy to produce one ion pair in air. Since there are 6.2×10^{18} ion pairs per coulomb of charge and 1 J of energy equals $1/(6.2 \times 10^{18}$ eV), then it takes 34 J of energy to generate 1 C of ions. Therefore, 1 R of exposure results in (2.58×10^{-4} C/kg) $\times$ 34 J/C of ions) = 0.0087 J of energy absorbed per kilogram of air.

The energy absorbed per mass of material is referred to as the absorbed dose. The units of absorbed dose are the rad (1 rd = 0.01 J/kg) or the current unit of the gray (1 Gy = 1 J/kg = 100 rd). Therefore, from the above calculation, 1 R of exposure gives the air a dose of 0.87 rd or 0.0087 Gy. A 1 R photon beam impinging on similar materials gives a similar dose and a 1 R photon beam results in a dose to tissue of approximately 1 rd or 0.01 Gy.

Since dose is usually measured at a particular location, it is sometimes useful for radiation safety purposes to determine an equivalent dose — the absorbed dose multiplied by some modifying factors. These factors include a radiation weighting factor for radiation such as neutrons or alpha particles, which are biologically more damaging for a given amount of energy absorbed, or tissue weighting factors, which attempt to determine an equivalent whole body dose when only portions of the body have been exposed. The equivalent dose is measured in units of Roentgen equivalent man (rem) and, more recently, in sieverts where 1 Sv = 100 rem.

Unit 23 summarizes the theory, instrumentation, and methods of monitoring applicable to some types of ionizing radiation.

(iv) Temperature Extremes

Heat and cold stress are physiological responses to extremes of hot and cold temperatures, respectively. Accordingly, thermal agents must be considered as part of the evaluation of the occupational environment. The body generates metabolic heat via basal and activity processes. If the environment is too cold, the body loses heat faster than it can produce heat, and a condition known as cold stress can develop. Alternatively, if the environment is too hot and the body is not able to cool fast enough, heat stress can result. The potential for heat stress is exacerbated if an individual is also very active and simultaneously exposed to elevated temperatures.

Body heat exchange is mainly accomplished by convection, radiation, and evaporation. The rate of heat exchange between the skin and the environment depends on different factors for each of these three processes. For convection, the skin temperature, air temperature, and wind velocity are important. For radiation, skin temperature and temperature of the surrounding objects determine the rate of heat exchange. For evaporation, humidity and wind velocity are the major determinants. When thermal equilibrium is no longer maintained under conditions of elevated temperatures, the result can be heat exhaustion, heat cramps, or heat stroke, depending on the duration and severity of stress and the effectiveness of steps taken to relieve the stress.

Heat stress is influenced by several environmental factors that include air temperature measured in degrees Celsius or Fahrenheit, air movement measured in feet per minute, humidity measured in percent moisture, and radiant heat also measured in degrees Celsius or Fahrenheit. Elevated air temperature increases the heat burden on the human body. The adverse influence of air temperature is exacerbated by the presence of excessive water vapor in the air due to high humidity. Elevated humidity decreases the evaporation rate of perspiration from the skin. Perspiration and its subsequent evaporation are natural defense mechanisms to transport excessive heat from the body. Excessive water vapor in the air hinders the evaporation process and the elimination of heat via evaporative cooling. The evaporative cooling is facilitated by increased air movement and, accordingly, hindered by lack of sufficient wind or breeze. Finally, radiant heat from various sources penetrates through the air and is absorbed by the body. In turn, the absorbed radiant heat increases the heat burden on the body. These thermal factors, in combination with metabolic factors related to workers based on their physiology and degree of exertion, influence the potential for developing heat stress. In addition, the potential for heat related disorders is increased by such factors as physical conditioning, age, weight, gender, and impermeable protective clothing.

Unit 19 summarizes the theory, instrumentation, and method of monitoring applicable to temperature extremes.

CHEMICAL AGENTS

Chemical agents are classified as inorganic and organic and are most noted for toxicity and flammability from an occupational health perspective. Although external exposure to chemical agents can occur via ingestion, dermal absorption, and injection, inhalation is considered the most common mode of entry in most occupational and even many nonoccupational environments. The inhalable forms of chemical agents are categorized as airborne aerosols (solid and liquid forms) and gases and vapors (gaseous forms).

Aerosols, also known as particulates, are divided into dusts, fibers, mists, fumes, and smoke. Dusts are generated from mechanical actions such as crushing, grinding, or sawing solid materials. Most dusts, and aerosols in general, have sphere-like or plate-like shapes. Fibers are dusts too, but tend to have more columnar rather than spherical or plate-like shapes characteristic of other dusts and aerosols. The columnar shape is characterized by a measurable length and width (diameter). Mists are finely divided liquid droplets generated via agitation of liquids or certain operations such as spraying or dipping. Mists are also generated by the condensation of gases and vapors. Fumes result when a metal is heated to its melting and sublimation points. The solid metal is eventually thermally converted to hot metal vapor. The hot vapor rises into the air where it often reacts with molecular oxygen to form metallic oxides. The vapor then condenses and forms very small sub-micrometer particles (fumes). Smoke particles are highly carbonaceous and formed from incomplete combustion reactions.

Aerosols are physically characterized by their shapes (e.g., sphere-like, disk-like, columnar) and sizes (e.g., aerodynamic equivalent diameter, length-to-width ratio). Aerosols can be separated from air via natural settling and via various anthropogenic modes, including filtration. Filtration involves a combination of diffusion, interception, and impaction of aerosols onto a solid, porous medium. All chemical aerosols, except fibers, are typically measured as concentrations of mass per volume in units of milligrams or micrograms per cubic meter of air (mg/m^3 or $\mu g/m^3$). Fibers are measured as counts per volume in units of fibers per cubic centimeter or cubic meter of air (f/cm^3 or f/m^3). Unlike gases and vapors, aerosols are not uniformly distributed throughout their area of generation.

Gases and vapors, besides those that comprise the atmosphere, are frequently generated or present in occupational and nonoccupational environments. True gases exist in the gaseous state at standard temperature (0°C or 273 K) and pressure (1 atm) (stp) and under conditions known as normal temperature (25°C or 298 K) and pressure (1 atm or 760 mmHg) (ntp). Sources of gaseous vapors, however, exist as liquids or even solids at stp and ntp. Vapors are generated from relatively volatile liquids and solids in relation to changes in pressure and temperature. Gases and vapors form true solutions in the atmosphere, which means that they expand and mix completely with ambient air. In relation, molar gas volumes of gaseous substances are approximately 22.5 l (lambert) at stp and 24.5 l at ntp.

Gaseous substances are not classified by size-like aerosols. Unlike aerosols, gases and vapors are of molecular size, and, accordingly, this means that they cannot be removed from the air via physical filtration methods. In addition, gases and vapors are monodisperse in terms of molecular size, meaning that the molecules of a particular gas or vapor will have homogenous size. The gas or vapor must somehow be separated from the surrounding air molecules via adsorption onto a solid or absorption into a liquid. Toxic gases and vapors are measured as concentrations of volume per volume in units of parts per million or per billion (ppm, ppb). Concentrations of flammable gases and vapors and oxygen gas are also measured as volume per volume, but in units of percent gas or vapor in air.

(i) Toxic Chemicals

Toxic substances include metals (e.g., elemental, salts, hydrides, oxides), aqueous-based acids and alkalines, and petroleum-based (e.g., aliphatic hydrocarbons, aromatic hydrocarbons, substituted hydrocarbons) solvents among others. The human toxicity of chemical agents is related to several factors including the duration of exposure, concentration, and mode of contact. Inhalation and dermal contact are the primary and secondary modes of exposure, respectively, to chemical agents in most occupational environments. Accordingly, the occupational environment is commonly evaluated for airborne toxic chemical agents. Exposure to chemical agents can therefore occur directly via dermal contact with the material (e.g., liquids, solids) during handling and indirectly via respiratory and dermal contact with airborne agents (e.g., aerosols, gases, vapors). Materials can also be ingested via handling food with contaminated hands or as a result of mucociliary escalation of materials that were originally inhaled into the respiratory system and subsequently transported from the lungs to the throat and mouth where they were swallowed.

Toxic chemical agents are either organic or inorganic. The organic materials are either hydrocarbons or substituted hydrocarbons. The hydrocarbons are composed solely of carbon and hydrogen, while substituted hydrocarbons are composed of carbon and hydrogen plus functional groups composed of elements such as chlorine, nitrogen, phosphorous, sulfur, oxygen, and so forth. The hazards associated with organic compounds and the fate of the compounds in the environment are mainly dependent on their chemical compositions and associated physical properties. Toxic inorganic materials are either metallic or nonmetallic elements, commonly in the form of salts, hydrides, and oxides.

Although distinctions are made between the subclasses of chemical agents, it should be noted that a single chemical component may exhibit a combination of characteristics. Indeed, in view of the fact that any chemical is toxic at an appropriate dose or concentration, then all flammable, corrosive, and reactive materials, as well as radiological and some biological agents, are also toxic. The reverse is not necessarily true. All toxic materials do not exhibit other hazardous characteristics.

Toxic agents may induce biochemical and physiological changes in human systems following either systemic contact, via absorpotion into blood and tissues, or local contact with respiratory, gastrointestinal, and dermal routes and systems. The changes may be ultimately manifested as adverse effects such as morphological and functional abnormalities, illnesses, and premature deaths among those exposed or their offspring. As suggested earlier, toxicity is inherent in all compounds. The toxicity of a given agent may be directly attributable to an original parent compound or indirectly attributable to an active metabolite formed via biotransformation in a human system. In addition, it also should be recognized that secondary toxicants can be generated by flammable, corrosive, and reactive chemicals in the form of toxic by-products released from reactions, fires, and explosions.

Units 4–10, 12, 13, and 14 summarize the theories, instrumentation, and methods of monitoring applicable to toxic chemical agents.

(ii) Flammable Chemicals

Flammable chemical agents include materials that serve as fuels (reducing agents) that can ignite and sustain a chain reaction when combined in a suitable ratio with oxygen (oxidizing agent) in the presence of an ignition source (e.g., heat, spark). Flammable agents are characterized as having a low flash point (i.e., <60°C), which is directly related to vapor pressure and, in turn, the volatility of a chemical substance. In general, organic compounds vaporize at relatively lower temperatures and, accordingly, are much more sensitive to heat than inorganic compounds. Flammable materials pose the obvious hazard of causing potential burns to human tissue. Indirectly during combustion, however, flammable materials can contribute to the formation of toxic atmospheres due to generation of by-products such as strong irritants and chemical asphyxiants, as well as via consumption of molecular oxygen during combustion.

Unit 11 summarizes the theory, instrumentation, and method of monitoring applicable to flammable chemical agents.

(iii) Corrosive Chemicals

Corrosive agents are materials that can dissolve metal in a relatively short period. Corrosive chemical agents also include those materials that can induce severe irritation and destruction due to accelerated dehydration reactions upon contact with human tissue. Typical examples of corrosives are organic and inorganic acids and bases. The strengths of acids and bases and the extremes of pH (i.e., <pH 2 and >pH 12.5, respectively) are correlated with the degree of corrosiveness.

Units 10 and 14 summarize the theory, instrumentation, and method of monitoring applicable to some corrosive chemical agents.

(iv) Reactive Chemicals

Reactive chemical agents consist of chemically unstable materials, which are typically characterized as either strong oxidizing or reducing agents. Chemical instability results in increased sensitivity to violent reactions that may result in extremely rapid generation of heat and gases. This, in turn, may culminate in ignition, explosion, or emission of toxic by-products. Some unstable compounds can react with air or water.

Units 4, 7–13, and 14 summarize the theories, instrumentation, and methods of monitoring applicable to some reactive chemical agents.

BIOLOGICAL AGENTS

Biological agents include pathogenic microorganisms. There are numerous identified pathogenic microbiologic agents that may be encountered in the environment. The pathogens consist of agents, which if introduced into the human body, may disrupt normal biochemical and physiological processes and activities via modes of infection, allergy, or toxicity. Infection is related mostly to the virulence and the population density of organisms present at a given target site in a host (i.e., human). Allergy is related to an exaggerated immune response as a result of a biological agent being recognized as a foreign body or antigen. Toxicity can be induced by some microbiological agents that synthesize chemical toxins (e.g., Gram negative bacteria endotoxins, fungi mycotoxins). The disruptions to humans that result from microbial-related infections, allergic reactions, or intoxications include acute and chronic illnesses and even death, if the host's immune system is unable to destroy the pathogens soon enough.

Examples of microbial pathogenic agents include bacteria, actinomycetes, rickettsia, fungi, protozoans, helminths, nematodes, and viruses. Unlike radiological and chemical substances, all biological agents, except viruses, are examples of biotic or living organisms. Viruses are abiotic or nonliving agents composed of biochemicals (i.e., protein-coated nucleic acids) that may insert into and disrupt human cells. Concentrations of microbiological agents are measured as:

- Counts per volume in units of colony forming units (CFUs) per cubic meter or liter (CFUs/m^3 or CFUs/l) for air samples (e.g., bioaerosols) and CFUs per liter (CFUs/l) for water and other liquid samples
- Counts per mass in units CFUs per kilogram or gram (CFUs/kg or CFUs/g) for soil and other solid samples
- Counts per area in units CFUs per square centimeter or square inch (CFUs/cm^2 or CFUs/in^2) for swabbed surface samples

Units 14 and 15 summarize the theories, instrumentation, and methods of monitoring applicable to biological agents.

ERGONOMIC FACTORS

Ergonomic factors include human attributes, abilities, and limitations as applied to the living and occupational environments. Ideally, optimal conditions will exist to maximize human health, comfort, and well-being, while promoting performance efficiency and effectiveness through the appropriate design of tools, machines, tasks, and environments.

Specific human factors include psychological capabilities, physiological dimensions (anthropometrics) and capabilities (biomechanics), and psychosocial issues. Fatigue; boredom; occupational stress; mental (cognitive) abilities and limitations; circadian rhythms; sensory capabilities; anthropometric and biomechanical attributes; and peer, labor-management, or organizational climate variables are just a few of the many factors that can contribute to unfavorable conditions in the occupational setting.

Some of the workplace factors of concern include hazards associated with the machines, equipment, tools, and layout and design variables found in the occupational environment. Some of the many variables of concern when assessing the causes of occupational-related ergonomic problems include:

- Confusing displays (gauges)
- Controls requiring excessive force for operation
- Hand tools requiring users to assume awkward body postures or positions
- Materials that are excessively heavy
- Manual materials handling with unreasonable frequency, duration, pace, or transfer distance requirements

Finally, physical, chemical, and biological agents are contributing factors from a holistic ergonomic perspective. Exposure to these environmental agents, including temperature extremes, humidity, lighting, noise, and air contaminants, can contribute to the ergonomic problems observed in the occupational setting. For example, excessively high workplace temperatures in conjunction with high levels of humidity can result in considerable discomfort and possibly health-related problems. Whereas, light and vision adaptation when moving from dark to well-lighted environments could contribute to safety-related problems.

Unit 24 summarizes the theory, instrumentation, and methods applicable to evaluating ergonomic factors.

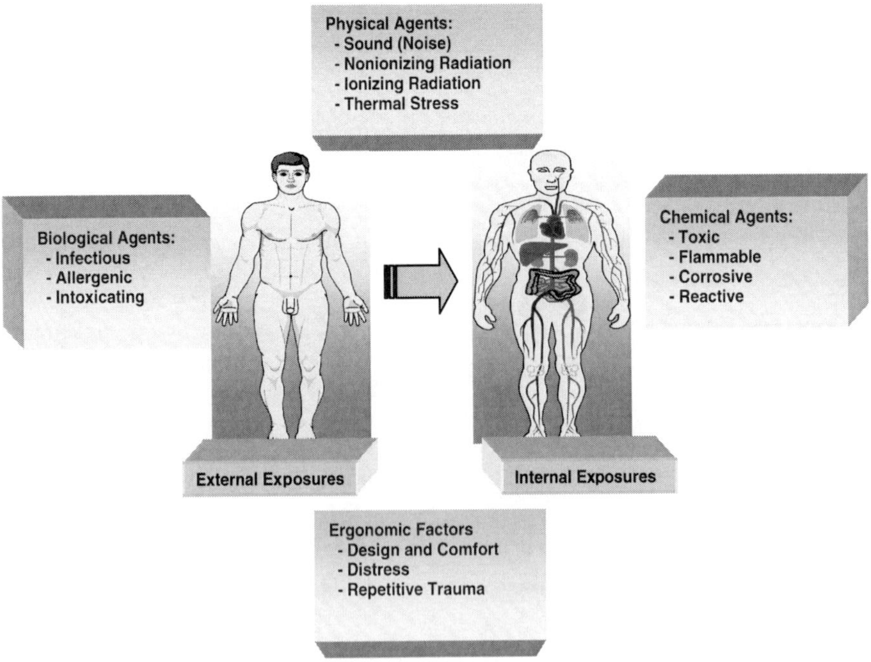

Figure 2.1 Exposures to environmental hazardous agents and factors.

UNIT 3

Sampling and Analytical Instruments Used to Evaluate the Occupational Environment: Generic Descriptions and Some Related Aspects of Calibration

LEARNING OBJECTIVES

At the completion of Unit 3, including sufficient reading and studying of this and related reference material, learners will be able to correctly:

- Describe the general categories and applications of major sampling and analytical instruments used for evaluating the occupational environment.
- Compare and contrast two major primary and secondary calibration standards used for measuring air sampling flow rates.
- Describe and explain the function of the frictionless bubble-tube flow meter.
- Describe and explain the function of the precision rotameter.
- Calibrate a rotameter against a frictionless bubble-tube flow meter.
- Develop a calibration curve for a precision rotameter.
- Calibrate an air sampling pump using either a primary or secondary calibration standard.
- Summarize the calibration procedure using primary or secondary calibration standards.

OVERVIEW

Various types and models of sampling and analytical instruments are used to evaluate the occupational and, indeed, some nonoccupational (e.g., homes) environments for potentially hazardous agents and factors. The instantaneous and integrated monitoring instruments are typically electronic and operate via direct current (DC power) from alkaline or rechargeable nickel-cadmium batteries. Some monitoring devices operate manually or even passively. For the most part, analytical devices all operate electronically via alternating current (AC power). Some monitoring instruments, including passive devices and related methods provide an analytical result instantaneously without the need for a separate phase of analysis. These are referred to as direct reading instruments and devices.

For all instruments, typically both before and after sampling and analysis are conducted, the devices are evaluated to determine and check if they are operating within defined qualitative and quantitative parameters. This evaluation is referred to as calibration. The evaluation includes a qualitative check of battery-operated electronic instruments to assure that the batteries are adequately charged for the intended type and duration of use. Some instruments, especially the direct reading type, are also qualitatively checked to determine if they detect agents or conditions for which they are designed and intended to respond. In relation, these instruments are quantitatively checked to determine if they are calibrated or adjusted to assure that the device measures agents detected within an acceptable range of concentrations or levels. Also, instruments involving active-flow pumps are quantitatively evaluated to assure that the flow rate (i.e., air volume moved per time) is calibrated to the desired setting.

Sampling and analytical instruments are needed to physically collect a sample, qualitatively detect and identify agents of interest and concern, and quantitatively measure the amount or level of the agents. Most integrated or continuous monitoring devices only collect the samples. The collected samples, in turn, are submitted to and analyzed at a laboratory where analytical instruments are used for detection, identification, and measurement of the agents. The direct reading instantaneous or real-time monitoring instruments, sample, detect, identify, and measure agents based on various types of combined collection and direct read-out technologies. Both integrated and instantaneous categories of instrumentation serve important purposes as part of comprehensive industrial hygiene and human exposure assessment practices.

DESCRIPTION OF SAMPLING AND ANALYTICAL INSTRUMENTS

A summary of some major sampling and analytical devices used to evaluate the occupational environment and some of the related aspects of calibration follows.

(i) Electronic Air Sampling Pumps for Conducting Integrated or Continuous Sampling of Particulates, Gases, and Vapors

Most integrated monitoring for airborne aerosols and particulates, gases, and vapors requires battery-operated air sampling pumps. The pumps serve as air moving devices to enhance active flow of air and airborne contaminants through a collection medium (e.g., filter, solid, or liquid sorbent) or into a collection container (e.g., plastic bag). Typical electronic active flow air sampling pumps used for industrial hygiene monitoring are categorized as low-flow (<1 l/min), high-flow (1 to 4 l/min), and ultrahigh-flow (>4 l/min). Refer to Figure 3.1. Multiflow air sampling pumps are popular since they can be used for either low- or high-flow applications. Most ultrahigh-flow pumps require energy from an alternating current (AC power) source (i.e., electrical outlet, electric generator) since the ultrahigh-flow rate for prolonged periods would drain the charge from batteries rather quickly. Nickel-cadmium (Ni-Cd), and more recently lithium-ion are examples of rechargeable batteries for air sampling pumps used for integrated sampling and operating meters used for instantaneous sampling.

The air sampling pumps are connected to a sampling medium or to a sampling container via a flexible hose. Air and airborne contaminants enter initially through the sampling medium or into the container for collection. Sampling media actually separate the contaminant from the airstream via filtration, adsorption, or absorption. The actual air, either without contaminant or with traces of contaminant that was not efficiently collected, actively flows from the medium into the flexible hose, serving as a conduit to transport the air to the pump. Ultimately, the air flows through and out of the pump. The sampling medium is submitted for laboratory analysis.

The two most common categories of sampling media for use with integrated air sampling pumps are filters and sorbents. Subcategories of filters include: membrane filters (e.g., polyvinyl chloride, mixed-cellulose ester), glass and quartz, and polycarbonate. Subcategories of sorbents include the solid a<u>d</u>sorbents activated charcoal and silica gel and various liquid a<u>b</u>sorbents. Filters and sorbents are discussed in more detail in the applicable units that follow.

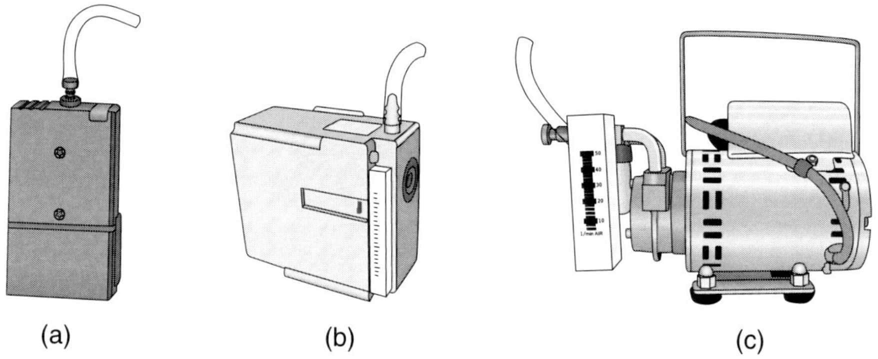

Figure 3.1 (a) Low-flow pump, (b) high-flow pump, and (c) ultrahigh-flow pump for conducting integrated or continuous sampling of particulate and gaseous agents.

Electronic air sampling pumps used for integrated monitoring are calibrated to assure that they operate at a known and accurate flow rate. The pumps are commonly calibrated using frictionless bubble-tubes or rotameter flow meters as described later in this unit.

Since the tubing and the presence of a sampling medium in-line can cause a loss in air pressure (i.e., pressure drop) and in turn airflow due to friction and turbulence, most air sampling pumps are currently designed to be constant flow pumps. The constant flow pumps respond to increased pressure drops by increasing the pumping mechanism. As a result, the airflow rate remains more constant throughout the sampling period, even as contaminant accumulates on a sample medium.

Units 4–7, 9, 10, and 15 in this book summarize theories, instrumentation and methods of sample collection that require the use of electronic air sampling pumps for conducting integrated or continuous monitoring of airborne radiological (e.g., ionizing radiation), chemical, and biological agents.

(ii) Electronic Air Sampling Pumps for Conducting Instantaneous or Real-Time Sampling of Particulates, Gases, and Vapors

Most instantaneous monitoring devices for sampling and analyzing particulate and gaseous agents contain a battery-operated pump (Figure 3.2). The batteries are commonly rechargeable, but some use disposable types. The purpose of the pump is to move the air and maintain active flow. The sampled air actively flows across a sensor or through a detector. This detector responds instantaneously to the presence of the agent and electronically indicates the concentration via a direct readout meter.

Instantaneous or real-time monitoring instruments for gases and vapors are calibrated against a known concentration of contaminant or reference gas to check the accuracy of response of the instrument. The known standards of chemical agents can be prepared and are also available commercially as compressed gases. Periodically, the electronic real-time monitoring devices for particulates, gases, and vapors must be returned for manufacturer or factory calibration.

Units 8, 11, and 13 summarize theories, instrumentation, and methods of sample collection that require the use of electronic air sampling pumps for conducting instantaneous or real-time monitoring of airborne chemical (e.g., aerosols, true gases, vapors) agents.

(iii) Manual Air Sampling Pumps for Sampling Gases and Vapors

Some air sampling pumps do not require electrical energy for operation. Instead, they are manually operated and referred to as manual pumps (Figure 3.3). Nonetheless, the pumps also serve as air moving devices and enhance flow of air and airborne contaminants through a sampling medium. The sampling medium reacts with the contaminant and reveals a direct reading of concentration.

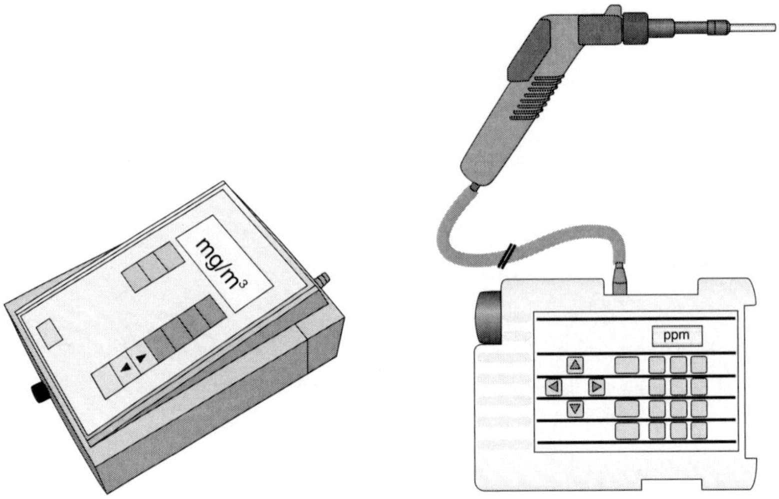

(a) (b)

Figure 3.2 (a) Total and respirable aerosol meter and (b) total ionizable organic gas and vapor meter for conducting instantaneous or real-time sampling of particulate and gaseous agents, respectively.

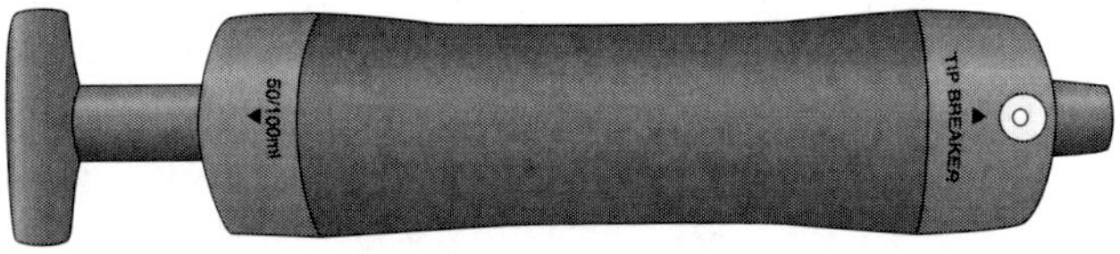

(a)

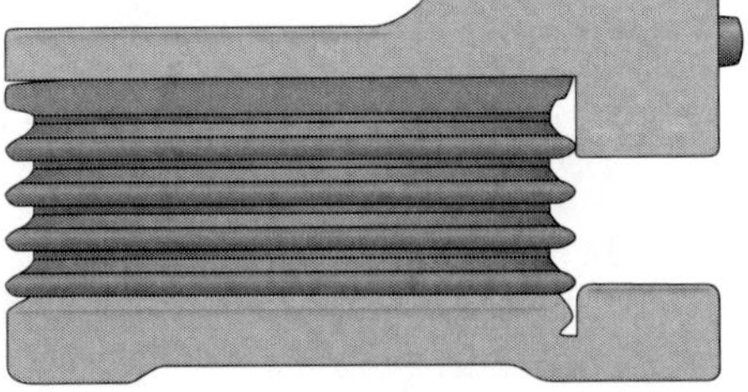

(b)

Figure 3.3 (a) Manual piston pump and (b) manual bellows pump for conducting instantaneous or real-time sampling of gaseous agents.

Two examples of manual air sampling pumps are the piston- and bellows-type pumps. These pumps are used with detector tubes to sample airborne gases and vapors and are commonly calibrated using a frictionless bubble-tube flow meter.

Unit 12 summarizes the theory, instrumentation, and methods of sample collection that require use of manual air sampling pumps for conducting instantaneous or real-time monitoring of airborne gaseous (i.e., true gases, vapors) chemical agents.

(iv) Electronic Meters for Sampling Energies

Many monitoring devices, especially those used for detecting and measuring physical agents or conditions, do not involve any type of pump or air moving device (Figure 3.4). The battery-operated devices, via the inherent electronics, sense and detect various forms of energies such as sound, vibration, temperature, ionizing radiation (e.g., alpha and beta particles) and nonionizing radiation (e.g., light; electric and magnetic fields; microwaves), and air velocity and flow (i.e., kinetic energy). Specific to this book and in common use for evaluating the occupational environment are sound level meters and octave band analyzers, audiometers, wet-bulb globe temperature meters, air velometers, ionizing radiation meters, illumination meters, electromagnetic field (EMF) meters, microwave meters, and spirometers/pulmonary function meters. The energy detected is converted to a direct-reading measurement.

Electronic meters for monitoring physical agents are typically calibrated against a known standardized source of energy compatible for the particular instrument. For example, electronic sound level meters are calibrated using a sound calibrator that generates a known level and frequency of sound. The sound calibrator is actually calibrated by the manufacturer. As with most electronic real-time monitoring devices, meters for detecting and measuring physical agents must be periodically factory calibrated by the manufacturer or the equivalent.

Units 16, 17, 19, 20, 21, 22, and 23 summarize theories, instrumentation, and methods of sample collection that require the use of electronic meters for conducting integrated and instantaneous monitoring of airborne physical (e.g., sound, nonionizing radiation) agents.

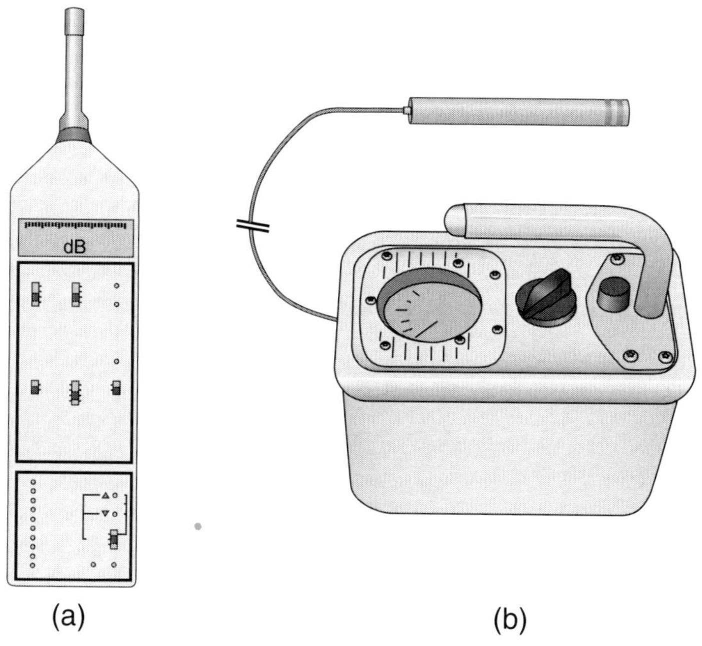

Figure 3.4 (a) Sound level meter-octave band analyzer and (b) ionizing radiation meter for conducting instantaneous or real-time sampling of physical agents (energies).

(v) Electronic Devices for Analyzing Collected Aerosols, Gases, and Vapors

Various analytical devices are used to detect and measure samples collected via integrated monitoring. The components and principle of operation for each are summarized in units contained later in this book. Some of the most common devices used to analyze samples collected in the occupational and the general environment are the electrobalance for gravimetric analysis of sampled particulates; gas chromatograph for chromatographic analysis of sampled organic gases and vapors; ultraviolet and visible light spectrophotometer for spectrophotometric analysis of sampled inorganic and organic gases, vapors, and mists; atomic absorption spectrometer and inductively coupled plasma emission spectrometer for spectroscopic analysis of sampled metal dusts and fumes; and phase contrast microscope for microscopic analysis of sampled fibers and organisms.

Electrobalances are factory calibrated, but can be checked for accuracy by weighing an object with known weight. Gas chromatographs and the various spectrophotometers are calibrated by running known concentrations of standard analytes to assure that the instrument responds or measures the analyte within an acceptable range. Calibration curves are established showing the relationship between the instrument response vs. a gradient of known amounts and concentrations of analytes. Microscopes and accessories (e.g., ocular graticule) are calibrated to assure accurate magnification and viewing area.

These analytical instruments and related methods are summarized or addressed in the following units in this book: electrobalance gravimetric analysis in Units 4 and 5; microscopic analysis in Units 6, 14, and 15; atomic absorption spectroscopic analysis in Units 7 and 14; gas chromatographic analysis in Units 9 and 14; and UV/visible spectrophotometric analysis in Units 10 and 14.

(vi) Miscellaneous Electronic and Manual Devices for Evaluating Meteorological Conditions

Important meteorological conditions should be measured and documented during most field and laboratory calibrations, sampling, and analyses. Both manual and electronic instruments are available for measuring each of the major parameters of temperature, pressure, and humidity. Air temperature (°C or °F) can be measured using a standard (dry bulb) thermometer. A sling psychrometer also can be used to measure air temperature based on the dry bulb temperature. The sling psychrometer consists of both dry bulb and wet bulb thermometers. The wet bulb thermometer is a standard thermometer with the bulb end covered by a cotton sleeve or wick moistened with deionized or distilled water. Following swinging or rotation of the sling psychrometer for about 1 min, the dry and wet bulb temperatures are read. The dry bulb reading alone indicates the air temperature. Comparison of both the dry and wet bulb temperatures to a special scale or psychrometric chart determines the corresponding percent relative humidity (% RH) in air. An electronic humidistat also can be used to measure % RH. Air pressure (mm Hg) can be measured using a barometer. There are some situations when air currents moving at substantial velocities warrant measurement. This is especially applicable when monitoring is conducted outdoors under relatively windy conditions. Air current or wind velocity (ft/min) can be measured using various types of velometers or anemometers.

SOURCES AND TYPES OF INFORMATION RELATIVE TO SAMPLE COLLECTION AND ANALYSIS

There are numerous sources of information relative to sample collection and analysis. Industrial hygiene monitoring is based on both science and art. It is not a simple cookbook technology where one simply reads the instructions for an instrument, turns it on, and begins sampling. There is much information that must be gathered to address monitoring before, during, and after it is conducted:

- What (i.e., agents), who (i.e., personnel), where (i.e., areas), and when (i.e., days, shifts) to evaluate?
- What are the processes and operations?
- Who (i.e., professional, technician) should conduct sampling?
- Which laboratory (accredited) should conduct the analysis?
- Which source (i.e., National Institute for Occupational Safety and Health [NIOSH], Occupational Safety and Health Administration [OSHA]) and method (i.e., name and number) for sampling and analysis?
- What is the sampling train, including media?
- What is the calibration?
- What are the flow rate and sampling duration (min-max sample volume)?
- What are the number of personnel and areas to evaluate and samples to collect, submit, and analyze?
- What is the number of field blanks to collect, submit, and analyze?
- What are the media storage, handling, and transport/shipping?

Guidelines and standards for selection of industrial hygiene and other applicable or related sampling and analytical instrumentation, methods, and operating procedures can be obtained from private, commercial, and governmental organizations and agencies. In addition to NIOSH and OSHA, these include, but are not limited to, the Mine Safety and Health Administration (MSHA), the American Conference for Governmental Industrial Hygienists (ACGIH), the American Industrial Hygiene Association (AIHA), the Environmental Protection Agency (EPA), the National Institute of Standards and Technology (NIST), the American Society for Testing and Materials (ASTM), the American National Standards Institute (ANSI), commercial and governmental laboratories, equipment and media vendors, and so forth.

CALIBRATION OF ELECTRONIC AND MANUAL AIR SAMPLING PUMPS

Air sampling pumps must be adjusted and calibrated to assure that the volume of air moved per unit time or flow rate during air sampling is accurate. Based on knowing the airflow rate of the pump and the duration of sampling, the total air volume sampled can be calculated. Air volume is the amount of air that literally flows into, over, and through a collection media or a collection chamber and sensor. Calculation of volume sampled is necessary so that the concentration of contaminant can be calculated. Concentrations are based on the amount of contaminant collected, detected, and measured in the sampling medium divided by the volume of air that flowed over, into, and through the sampling medium.

$$
\begin{aligned}
Concentration &= \frac{Amount}{Air\ Volume} \\
&= \frac{Aerosol\ Weight}{Air\ Volume} \\
or, &= \frac{Fiber\ Count}{Air\ Volume} \\
or, &= \frac{Volume\ Gas}{Air\ Volume} \\
or, &= \frac{Colonies}{Air\ Volume}
\end{aligned}
\tag{3.1}
$$

Air sampling pumps should be calibrated both pre- and post-sampling. Calibration involves adjusting pumps to a specific flow rate and measuring against a reference or standard device (i.e., calibrator). Flow rates usually are expressed in units of liters per minute (l/min) and cubic centimeters (cm^3) or milliliters (ml) per minute (1 cm^3/min = 1 ml/min). Ideally, the average flow rates measured during pre- and post-sampling calibration should be within ±5% of each other.

The most fundamental calibration methods for most practicing industrial hygienists involves the use of primary and secondary calibration devices to set and document the flow rates of air sampling pumps. Accordingly, only two major types of calibration devices will be described in this book. Primary standards are the most accurate (≤1%) and commonly used calibration devices. Secondary calibration standards for flow are calibrated against primary standards.

The frictionless bubble-tube (or soap film) flow meter is an example of a major primary calibration standard commonly used for industrial hygiene applications (Figure 3.5). The bubble-tube consists of a transparent glass volumetric cylinder. A soap solution introduced at the bottom of the cylinder is used to generate a flat, circular disk-shaped bubble/soap film within the bubble-tube. An air sampling pump is connected to the top outlet of the bubble-tube via a length of flexible hose (Figure 3.6). The volume displacement of air per unit time, or flow rate, can be determined by measuring the time required for the soap film contained in the tube to move from an initial (e.g., 0 cm^3) to final (e.g., 1000 cm^3) scale markings that enclose a known volume (e.g., 1000 cm^3). The flow rate (Q [l/min]) is determined by dividing the volume traversed by the bubble (V [cm^3]) in the bubble-tube by the time measured (T [sec]) for the bubble to travel across the specific volume.

$$Q\,(1/\min) = \frac{V\,(cm^3)}{T\,(\sec)} \times \frac{60\ \sec}{1\ \min} \times \frac{1\ l}{10^3\ cm^3} \qquad (3.2)$$

Air temperature and atmospheric pressure should be measured during calibration and sampling procedures. The data can be used, for example, to adjust actual flow rate (Q_{act}) to standard flow

Figure 3.5 Manual frictionless bubble-tube calibrator used as a primary standard for calibrating air sampling pumps.

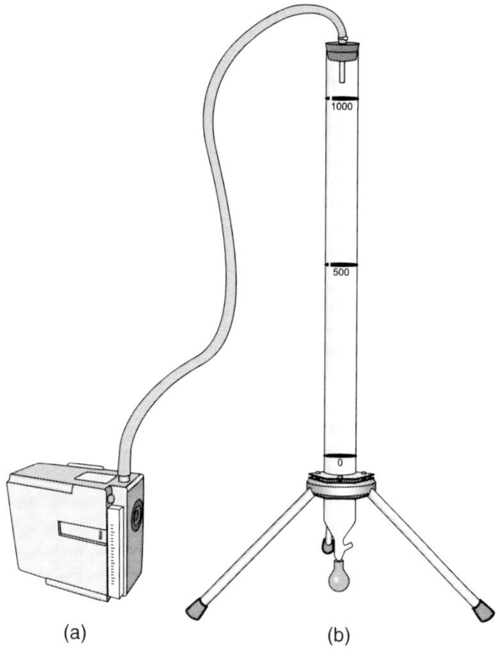

Figure 3.6 (a) High-flow pump connected to (b) a manual frictionless bubble-tube calibrator.

rate (Q_{std}). Although standard temperature is 0°C (273 K), industrial hygiene practice more often references a different standard or normal temperature of 25°C (298 K). Both standard and normal pressures are 760 mmHg.

$$Q_{std} = Q_{act} \times \frac{298}{T} \times \frac{P}{760} \tag{3.3}$$

Manual and electronic frictionless bubble-tube calibrators are commonly used. The manual devices require using a stopwatch to time the bubble as it traverses a known volume followed by calculations to determine the flow rate. Electronic bubble-tube flow meters have an electronic flow sensor (e.g., infrared) that automatically measures the time for the bubble to travel a given volume and automatically calculate the flow rate via a microprocessor and show an LCD display in liters per minute or milliliters per minute. A soap solution is used with both the manual and electronic frictionless bubble-tubes to form the soap film. An alternative to the electronic soap bubble-tube involves a dry-tube calibrator that automatically senses and times the movement of a low-frictionless weight (e.g., graphite/carbon) within a cylinder instead of the movement of a soap film. All electronic calibrators must be periodically factory calibrated and checked against another primary standard (Figure 3.7).

Electronic calibrators are rapidly replacing the manual frictionless bubble-tube calibrators for various reasons including compact size, ease and rate of use, and accuracy and precision. Nonetheless, novice learners of fundamental aspects of calibration should be familiar with use, principles, and calculations of the manual frictionless bubble-tube calibrators for a more tangible comprehension of measuring flow rate.

A rotameter is a major secondary standard for calibrating air sampling pumps (Figure 3.8). The device consists of a tapered, precision-bored tube, made of transparent glass or plastic, with a solid float inside. As with the bubble-tube calibrators, an air sampling pump is connected to the top outlet of the bubble-tube via a length of flexible hose (Figure 3.9). Air flows into the bottom of the tube and carries the float upward until a level is reached where the force of the air is offset by the weight

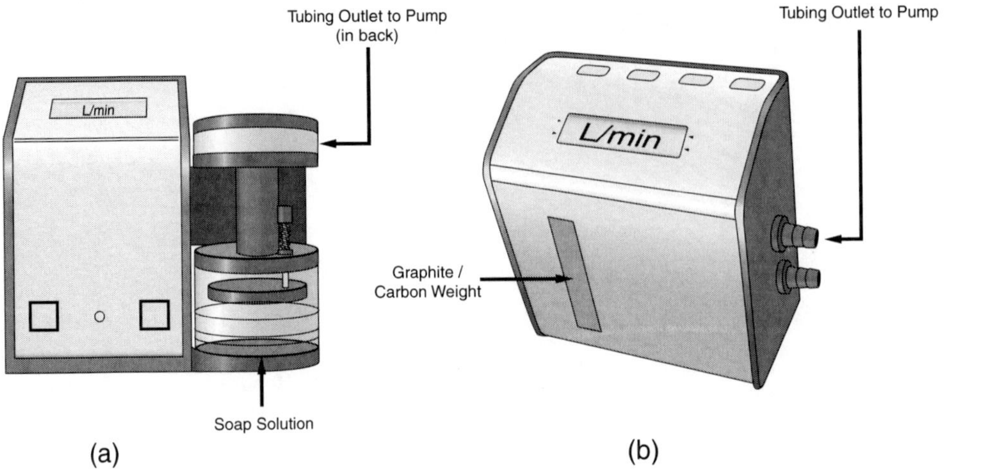

Figure 3.7 (a) Electronic frictionless bubble-tube calibrator and (b) electronic dry calibrator used as primary standards for calibrating air sampling pumps.

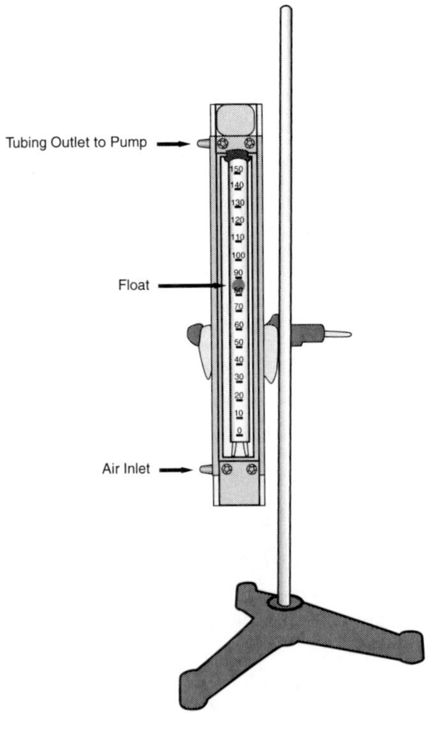

Figure 3.8 Rotameter.

of the float. The height of the float from the bottom of the tube represents the airflow rate. The typical corresponding rotameter reading is the number where the float is its highest, widest point.

Rotameters are calibrated by comparison with a primary calibrating device (Figure 3.10). For example, an active flow air sampling pump, rotameter, and bubble-tube are connected in-line using flexible hose. The flow rate readings measured using the bubble-tube and corresponding rotameter readings for perhaps five different flow settings on the pump are recorded. Data can be plotted using a linear graph showing rotameter readings vs. bubble-tube flow rate readings (Figure 3.11). Corresponding flow rates for various rotameter readings can later be extrapolated from the curve.

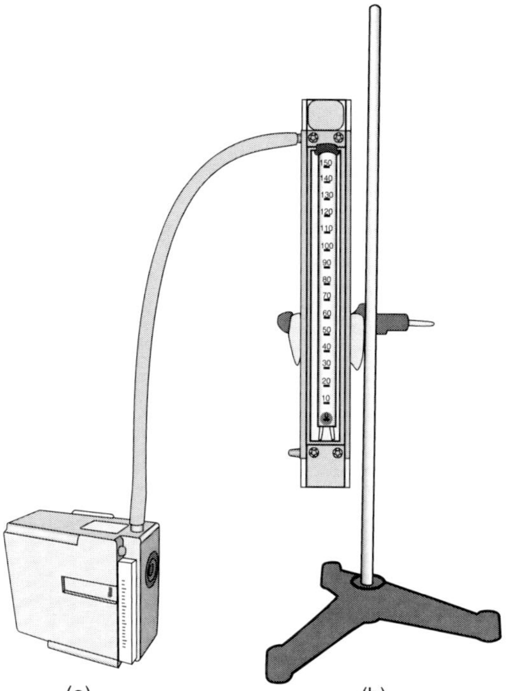

(a) (b)

Figure 3.9 (a) High-flow pump connected to (b) a rotameter calibrator.

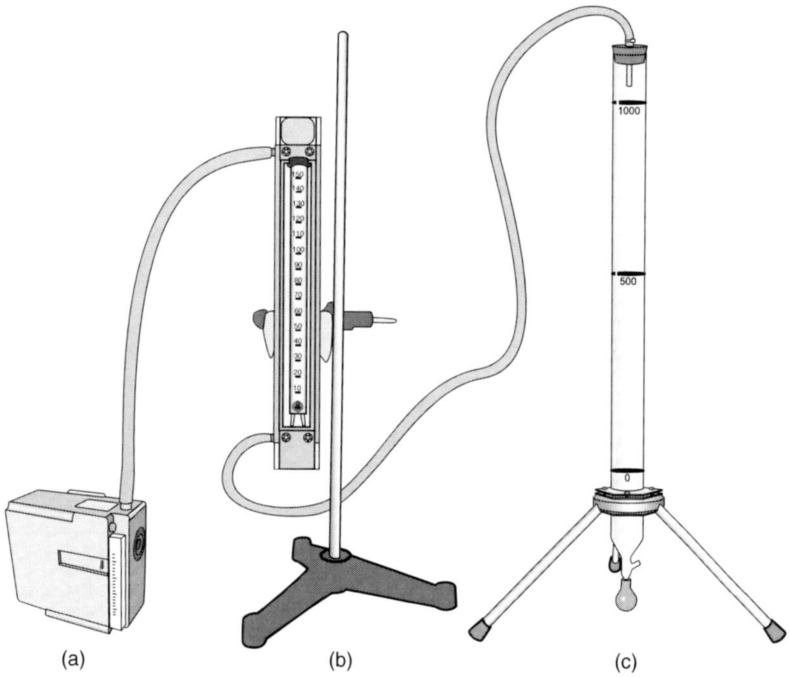

(a) (b) (c)

Figure 3.10 (a) High-flow pump connected to (b) a rotameter calibrator in-line with (c) a frictionless bubble-tube calibrator.

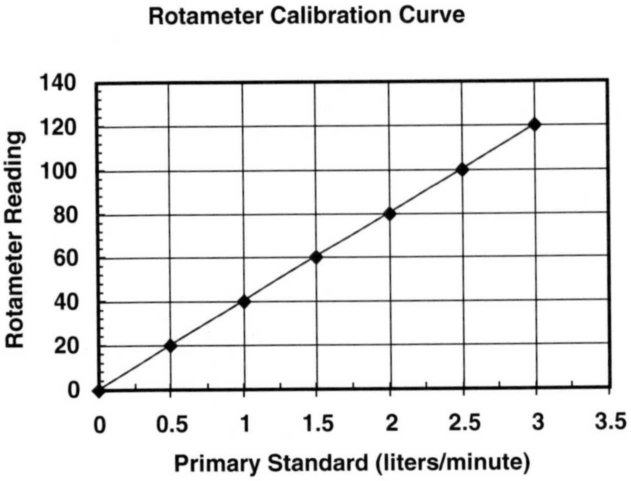

Figure 3.11 Rotameter calibration curve.

A rotameter should always be calibrated under similar conditions of pressure and temperature as it is to be used. Once calibrated, the rotameter can be directly used to calibrate air sampling pumps. This is especially useful in the field due to portability and simplicity. If a rotameter is calibrated at one location (Q_{cal} measured at T_{cal} absolute and P_{cal}) and used under conditions (T_{samp} absolute and P_{samp}) at the sampling location that differ considerably (i.e., >5%)[1], the equation shown below can be applied to correct for and calculate the equivalent (actual) flow rate. This correction only applies to calibrations using rotameters[2].

$$Q_{actual} = Q_{cal} \times \sqrt{\left(\frac{T_{samp}}{T_{cal}} \times \frac{P_{cal}}{P_{samp}}\right)} \qquad (3.4)$$

The portability of the rotameter and the electronic calibrators increases the feasibility of ideally calibrating air sampling pumps at the sampling location and periodically checking flow rates during sampling. However, although rotameters are an acceptable alternative to primary standard calibrators, preference should be given to the more accurate (±<1%), relatively compact, and versatile (5 ml to 30 l range) electronic bubble- or dry-tube calibrators discussed above.

UNIT 3 EXERCISE

OVERVIEW

The exercise will provide the fundamental concepts for calibrating an air sampling pump using a manual or electronic frictionless bubble-tube flow meter as a primary calibration standard. In

[1] Waldron, P.F., Principles and instrumentation for calibrating air sampling equipment, in *The Occupational Environment — Its Evaluation and Control,* DiNardi, S., Ed., AIHA Press, Fairfax, VA, 1997.

[2] McCammon, C.S. and Woebkenberg, M.L., General considerations for sampling airborne contaminants, in *NIOSH Manual of Analytical Methods (NMAM),* 4th ed., Cassinelli, M.E. and O'Connor, P.F., Eds., Department of Health and Human Services and National Institute for Occupational Safety and Health Publication No. 94-113, U.S. Government Printing Office, Washington, D.C., 1994.

addition, the exercise will demonstrate the calibration of a rotameter against the primary standard. Place a check mark (✔) in the open box (☐) when you have obtained applicable material and completed the steps for the calibration methods.

MATERIAL

1. Calibration of an Air Sampling Pump Using a Primary Standard

- ☐ Multi- or high-flow air sampling pump
- ☐ 2-ft length of 1/4-in. internal diameter (i.d.) flexible hose
- ☐ Manual or electronic frictionless bubble-tube flow meter (primary calibration standard)
- ☐ Soap solution
- ☐ Stopwatch (if manual calibrator is used)
- ☐ Bubble-tube calibration data form (Figure 3.12)

2. Calibration of a Secondary Standard Against a Primary Standard

- ☐ Same material listed in #1 above
- ☐ Rotameter (secondary calibration standard)
- ☐ 1-ft length of 1/4-in. i.d. flexible hose
- ☐ Rotameter calibration form (Figure 3.13)

METHOD

1. Calibration of an Air Sampling Pump Using a Primary Standard

- ☐ Remove the pump from charger, turn "ON," and allow it to operate for 5 min prior to calibrating.
- ☐ Attach one end of the flexible hose to the inlet port of the air sampling pump.
- ☐ Connect the other end of the flexible hose to a manual or electronic frictionless bubble-tube calibration device; this constitutes the air sampling pump calibration train (Figure 3.6).
- ☐ Aspirate the bubble solution into the calibration device to form a soap film (bubble).
- ☐ Record the time (T [sec]) it takes the soap film (bubble) to traverse a volume of 1000 ml.
- ☐ Calculate the flow rate (Q) in liters per minute.
- ☐ Complete a calibration data form.

2. Calibration of a Secondary Standard Against a Primary Standard

- ☐ Remove the pump from charger, turn "ON," and allow it to operate for 5 min prior to calibrating.
- ☐ Attach one end of the 2-ft length of flexible hose to the inlet port of the air sampling pump.
- ☐ Connect the other end of the flexible hose from the pump to the outlet port of the rotameter.
- ☐ Connect one end of the 1-ft length of flexible hose to the inlet port of the rotameter.
- ☐ Connect the remaining end of the flexible hose to a manual or electronic frictionless bubble-tube calibration device; this constitutes the rotameter calibration train (Figure 3.10).
- ☐ Adjust the flow control of the air sampling pump to the highest position.
- ☐ Allow the rotameter float to stabilize.
- ☐ Aspirate the bubble solution into the calibration device to form a soap film (bubble).
- ☐ Record the time (T [sec]) it takes the soap film (bubble) to traverse a volume of 1000 ml and record this value as A_1. Simultaneously, take a rotameter reading from the highest, widest point on the float and record this value as B_1.
- ☐ Continue the above, adjusting the flow control after each data pair (A and B) is recorded, so that the flow rate decreases by about one-fifth each time. Stop after five data pairs have been recorded.

Calibration Data Form:
Low-Flow, High-Flow, and Multi-Flow Air Sampling Pumps

Name and Location of Calibration: ____________________

Calibration Conducted By: ____________________ **Date Calibration Conducted:** ____________
Calibration Instrument
(Type/Manufacturer/Model): ____________________

Air Sampling Pump
(Type/Manufacturer/Model): ____________________

Collection Medium In Line: ____________________ **Calibrator Volume:** ________cc

Air Temperature: ____°C **Air Pressure:** ____mm Hg **Relative Humidity:** ____%

Pump No.	Pre-Sampling Calibration Flow Rate (Q_{pre})					Post-Sampling Calibration Flow Rate (Q_{post})					Average Flow Rate
	Time (sec)				Q_{pre}	Time (sec)				Q_{post}	Q_{avg}
	T_1	T_2	T_3	T_{avg}	/min	T_1	T_2	T_3	T_{avg}	/min	/min

Calibration Notes:

Figure 3.12 Calibration data form for air sampling pumps.

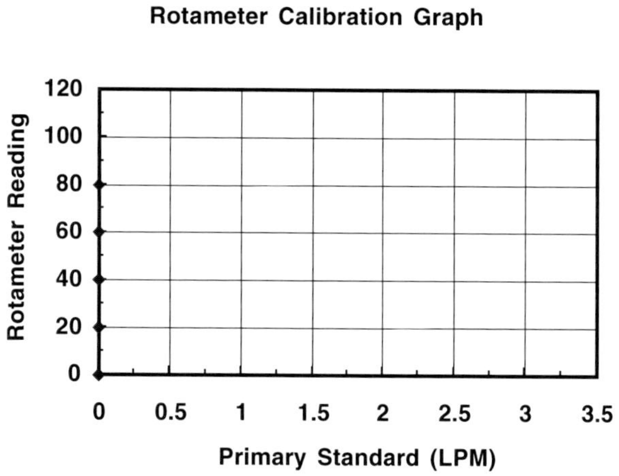

Figure 3.13 Calibration data form for rotameter.

- ☐ Draw an X-Y line graph on the rotameter calibration form. Label the Y-axis rotameter reading and the X-axis bubble-tube reading (l/min), plot the data pairs (A_1 B_1, A_2 B_2, A_3 B_3, A_4 B_4, A_5 B_5), and draw a straight line through the five coordinates.
- ☐ Disconnect the bubble-tube from the rotameter, leaving just the pump connected to the rotameter (Figure 3.9 and pressure [1 atm or 760 mmHg] [STP]).
- ☐ Arbitrarily adjust the flow control on the air sampling pump to a slightly higher flow rate and record the reading.
- ☐ Locate the corresponding reading from the Y-axis of your graph. Draw a horizontal line from that point and parallel to the X-axis until it intersects the linear curve, and then extend the line from the point of intersection and parallel to the Y-axis until it intersects the X-axis. Read the corresponding flow rate in liters per minute.

UNIT 3 EXAMPLES

Example 3.1

A multiflow air sampling pump was calibrated using a manual frictionless bubble-tube. What was the average flow rate (Q) in liters per minute (l/min) if 3 trials (n = 3) were run resulting in 25.5 sec (T_1), 25.2 sec (T_2), and 25.9 sec (T_3) for the bubble to traverse a volume of 1000 cm^3?

Solution 3.1

$$Q_{avg}\ (\mathrm{l/min}) = \frac{\mathrm{V}\ (cm^3)}{(T_1\ [\mathrm{sec}] + T_2\ [\mathrm{sec}] + T_3\ [\mathrm{sec}])/3} \times \frac{60\ \mathrm{sec}}{\mathrm{min}} \times \frac{1\ l}{10^3\ \mathrm{cm}^3}$$

$$= \frac{1000\ \mathrm{cm}^3}{(25.5\ \mathrm{sec} + 25.2\ \mathrm{sec} + 25.9\ \mathrm{sec})/3} \times \frac{60\ \mathrm{sec}}{\mathrm{min}} \times \frac{1\ l}{10^3\ \mathrm{cm}^3}$$

$$= 2.35\ \mathrm{l/min}$$

Example 3.2

A rotameter was calibrated using a frictionless bubble-tube as a primary standard. Data are shown below. What would the calibration curve look like? What is the flow rate (l/min) if the rotameter reading is 35?

Rotameter Reading	Bubble-Tube Flow Rate (l/min)
63	3.00
50	2.50
42	2.00
31	1.50
22	1.00
9	0.50
0	0

Solution 3.2

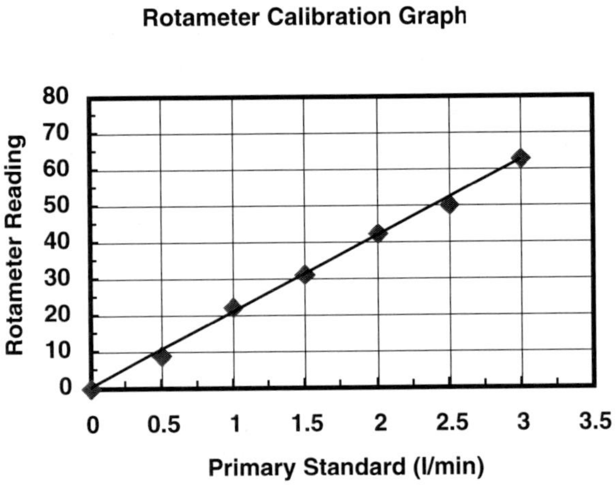

UNIT 4

Evaluation of Airborne Total Particulate: Integrated Personal and Area Monitoring Using an Air Sampling Pump with a Polyvinyl Chloride Filter Medium

LEARNING OBJECTIVES

At the completion of Unit 4, including sufficient reading and studying of this and related reference material, learners will be able to correctly:

- Name, identify, and assemble the components of a common sampling train, including the specific sampling medium, used for conducting integrated sampling of airborne total nuisance dust or particulate not otherwise classified, hereafter referred to as total particulate.
- Name, identify, and assemble the applicable calibration train that uses a primary standard for measuring flow rate of a multi- or high-flow air sampling pump.
- Calibrate a multi- or high-flow air sampling pump, with the applicable representative sampling medium in-line, using a mechanical or electronic frictionless bubble-tube.
- Summarize the principles of sample collection of airborne total particulate.
- Conduct integrated sampling of airborne total particulate.
- Prepare samples for analysis of collected particulate using gravimetry.
- Name and identify the components of a mechanical or electrical balance.
- Name, identify, and assemble the components necessary for analysis of samples for total particulates using an electrobalance.
- Check the calibration of an electrobalance using standard weights.
- Summarize the principles of sample analysis of filters used for collection of airborne total particulate.
- Conduct analysis of a sample for total particulate using an electrobalance.
- Perform applicable calculations and conversions related to pre- and post-sampling flow rate of an air sampling pump, average flow rate of an air sampling pump, pre- and post-weight of a filter medium, sampling time, sampled air volume, weight of total particulate collected, and concentration of total particulate in units of milligrams per cubic meter (mg/m^3).
- Record all applicable calibration, sampling, and analytical data using calibration, field sampling, and laboratory analysis data forms.

OVERVIEW

Occupational exposure limits, such as American Conference of Governmental Industrial Hygienists threshold limit values (ACGIH-TLVs) and Occupational Safety and Health Administration permissible exposure limits (OSHA-PELs), for airborne particulates have been established for many individual elements and compounds based on their inherent toxicities. There are situations when it is important to know what the concentration of particulate, both highly toxic and practically nontoxic, is in a given area. It is equally important to determine what an individual's external exposure is to total particulate rather than a specific type. Situations also exist in which there are no occupational exposure limits for a specific type or mixture of particulate air contaminants. These materials may be generically classified and sampled and analyzed as total dust or particulate. In either situation, measurements of total particulate, that is, particulate mixtures of various sizes and compositions, are frequently warranted. There is an OSHA-PEL for total dust and an ACGIH-TLV for particulates not otherwise specified. Sampling and analytical methods are available for these particulates, and involve moving an airstream into and through a filter collection medium. Airborne particulates are electrostatically attracted to, intercepted by, and impacted on the filter medium, resulting in separation of the particulates from the air without regard to particle size or composition (Figure 4.1).

The ACGIH also has TLVs based on size-selective sampling of three categories of particulate mass fractions[1]. These three categories of TLVs are for materials deemed hazardous when they deposit:

- Anywhere in the respiratory tract (inhalable particulate mass-TLV)
- Anywhere within the lung airways and the gas-exchange region (thoracic particulate mass-TLV)
- Anywhere in the gas-exchange region (respirable particulate mass-TLV)

Size-selective instrumentation and methods are needed and have been established to sample, analyze, and characterize airborne particulate relative to these categories. Some size-selective samplers are summarized in Unit 5.

FILTER MEDIUM FOR TOTAL PARTICULATE

The sample medium is a 37-mm diameter polyvinyl chloride (PVC) filter with a 5.0-μm pore size. The diameter (37 mm) of the filter provides cross-sectional surface area for deposition and collection of particulate. The pore size (5 μm) causes particulates with diameters greater than the pore diameter to collect on the surface of the filter. Particulates with diameters less than the pore size are also collected. These particles are collected within the pores due to an electrostatic attraction between the particles and the filter. The fine particulate is collected primarily via diffusion into the filter material whereas the relatively coarser particulate is intercepted and impacted within the medium. Collection efficiency increases as particulates accumulate on the filter. The larger pore size (5 μm) filter is used relative to smaller pore sizes (0.45 to 0.8 μm) to reduce clogging and overloading of the filter since total particulate is composed of diverse particle sizes.

Figure 4.1 Separation of particulate sample from air via filtration.

[1] The American Conference of Governmental Industrial Hygienists (ACGIH), *2002 TLVs and BEIs — Threshold Limit Values for Chemical Substances and Physical Agents and Biological Exposure Indices,* Cincinnati, OH, 2002.

Since sample preparation and analysis involves a weighing or gravimetric procedure, filters are stored in a dessicator until use. Standard dessicators contain a hygroscopic medium such as silica gel crystals. The purpose of dessication is to absorb any water that is present on the filter that would result in an erroneously higher weight when the filter is weighed pre- and post-sampling. PVC plastic filters are used instead of a paper cellulose type since the plastic is relatively hydrophobic. Nonetheless, dessication helps assure that the pre-weight is due only to the filter, and post-weight is due only to a combination of the filter and collected particulate. Filters should be dessicated for approximately 24 h prior to weighing. Dessication can be accelerated and incubation time reduced to less than 30 min if a vacuum dessicator is used.

The filter medium is positioned on a cellulose support pad and secured within a 37-mm three-stage plastic cassette (Figure 4.2). The bottom or outlet stage holds the porous support pad and the filter. It has an orifice (hole) in the center that serves as the outlet port for sampled air to exit the cassette. The middle stage is a ring-shaped spacer or cowl that increases the depth or height of the cassette to facilitate more even deposition of the particulates from the air as it enters the cassette. The top or inlet stage contains a hole in the center that serves as the inlet port for sampled air to enter the cassette.

PRECAUTIONS

Filters can be overloaded with particulate if airborne concentrations and sample volumes are too high. Particulates can visibly accumulate and form a mound or inverted cone in the center of the filter directly below the inlet orifice of the cassette. The accumulated particulate can break off and be lost during sample analysis if the cassette and sample are not handled carefully. Filters can also become contaminated with particulate if a dusty cassette is not wiped off prior to disassembling for analysis. In either case, loss of particulate from the filter or contamination of the filter can result in erroneous data and an invalid sample. These precautions are applicable for samples collected for any airborne particulate, including total dust, respirable dust, fibers, and metal dust and fume.

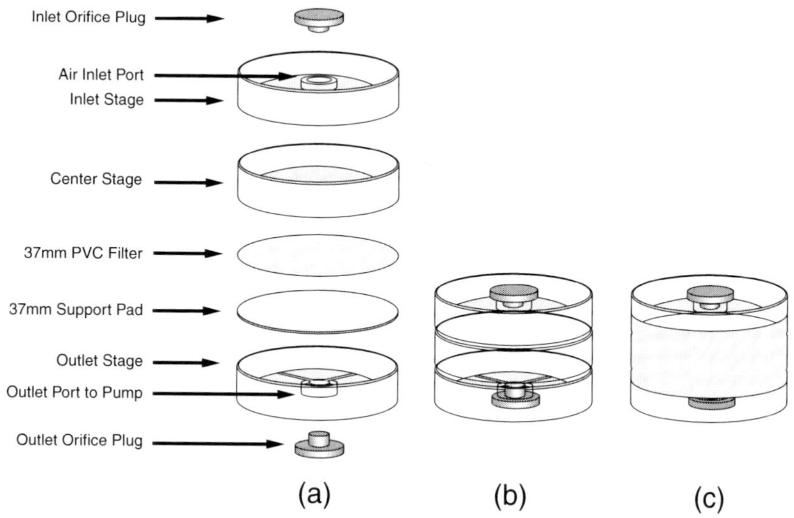

Figure 4.2 A 37-mm three-stage cassette with a 37-mm PVC filter and support pad. (a) Exploded view, (b) assembled view, and (c) assembled view secured with shrink band or tape.

SAMPLING

The sample collection medium is prepared by pre-weighing a filter for each field sample and blank that is going to be collected and ultimately submitted for laboratory analysis. Immediately prior to pre-weighing, the dessicated filter is passed across an ionization device to eliminate any static charge that may be present on the filter, and cause contamination due to an unwanted attraction of airborne particulate. The anti-static device contains a radioisotope such as polonium-210. A filter is pre-weighed using an electrobalance to determine pre-weight. The pre-weighed filter is subsequently positioned on a cellulose support pad in a three-stage plastic cassette. The cassette is assembled, and plugs are inserted at the inlet and outlet ports or orifices. When used for sampling, the orifice plugs are removed and the cassette is attached to an air sampling pump using a length of 1/4 in. internal diameter (i.d.) flexible hose (Figure 4.3). Typical sampling rates range from 1.5 to 2.0 l/min. Following sampling, the cassette is removed and plugs are reinserted in the open orifices in the field. The sample is then transported to a laboratory for analysis.

A brief outlined summary of sample collection follows.

(i) Calibration

- The air sampling pump is calibrated pre- and post-sampling to adjust or determine flow rate using a manual or electronic calibrator with representative media (i.e., an assembled three-stage cassette containing a 37-mm, 5.0-µm pore size PVC filter and support pad) in-line (Figure 4.4).

- Pre- (Q_{pre}) and post-sampling (Q_{post}) flow rates are determined by measuring the average time (T_{avg} sec) based on three trials ($\Sigma T_{1,2,3}/3$) for a bubble to traverse a specific volume (V cc) of the bubble-tube.

$$Q\,(\mathrm{l/min}) = \frac{\mathrm{V\,(cm^3)}}{T_{avg}\,(\mathrm{sec})} \times \frac{60\ \mathrm{sec}}{1\ \mathrm{min}} \times \frac{1\,\mathrm{l}}{10^3\ \mathrm{cm}^3} \tag{4.1}$$

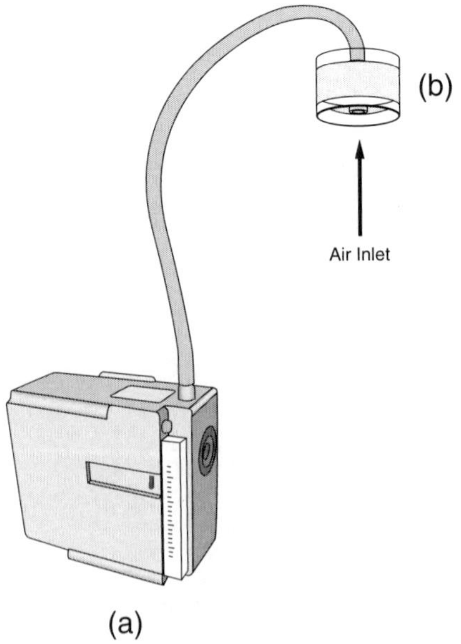

Figure 4.3 Sampling train for total particulate composed of (a) a high-flow pump connected with flexible hose to (b) a 37-mm PVC filter in three-stage cassette.

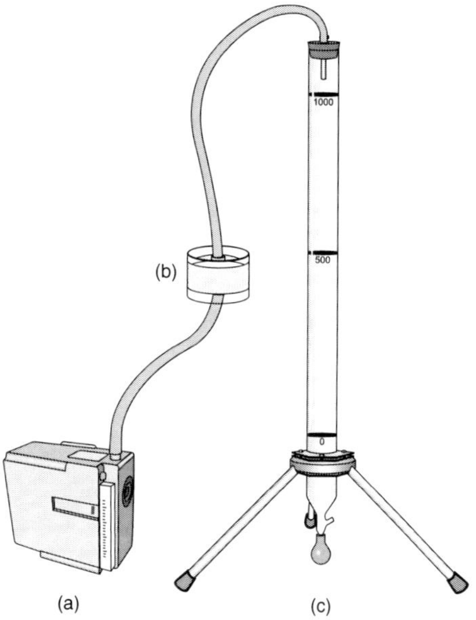

Figure 4.4 Calibration train for total particulate composed of (a) high-flow pump connected with flexible hose to (b) a 37-mm PVC filter in three-stage cassette in-line with (c) a frictionless bubble-tube.

- Average flow rate (Q_{avg}) is based on the average of pre- (Q_{pre}) and post-sampling (Q_{post}) flow rates.

$$Q_{avg} = \frac{Q_{pre} + Q_{post}}{2} \tag{4.2}$$

- Typical flow rates for sampling total particulate are 1.5 to 2 l/min.

(ii) Preparation for Sampling

- A weighed PVC filter plus unweighed support pad are inserted into a plastic cassette which is then assembled and labeled.
- The sampling train is assembled and consists of a multi- or high-flow pump; ¼ in. i.d. flexible hose with clips; and a 37-mm three-stage plastic cassette containing a weighed 37-mm, 5.0-µm pore size filter on an unweighed cellulose support pad.

(iii) Conducting Sampling

- A labeled plastic cassette is positioned with the inlet port facing downward and attached within the breathing zone. An air sampling pump connected to the cassette is clipped to the belt of the worker for personal sampling. The sampling train is positioned and secured in a specific location for area sampling.
- The pump is turned "ON" and start time is recorded.
- After a specified time, the pump is turned "OFF" and stop time is recorded.
- The orifice plugs are inserted into the inlet and outlet ports of the plastic cassette and the sample is transported to laboratory for analysis.

ANALYSIS

The sampling method for total particulate is similar to those methods used to detect and measure specific particulate air contaminants. The most significant difference is the method of analysis. Whereas specific particulate air contaminants require an analytical method that will both identify

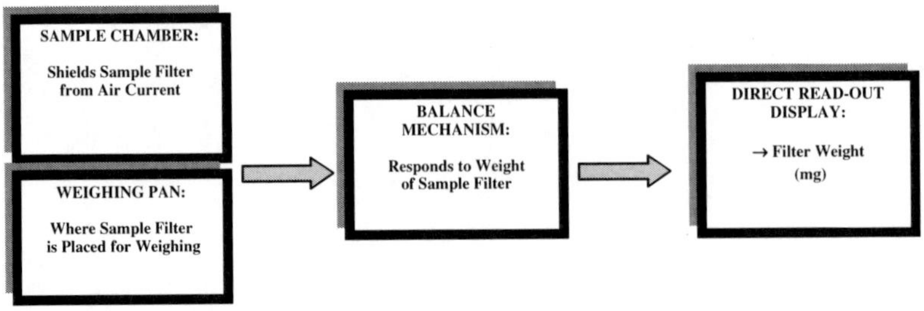

Figure 4.5 Components of a mechanical or electrical balance.

the specific analyte and measure the amount collected, analysis of total particulate only involves measurement of the amount (weight) collected via a simple gravimetric procedure. Accordingly, the identification of the specific components or composition of the collected particulate matter is not determined. Analysis of the sample involves a gravimetric method using a mechanical or electrical balance (Figure 4.5). The gravimetric procedure involves comparison of post- and pre-weight of filter. Theoretically, the pre-weight is the weight of only the filter and the post-weight is the weight of only the filter plus the collected particulate. The difference in weights (post-sample minus pre-sample filter weight) represents the total particulate collected. Gravimetric analysis is the most simplistic of the major analytical methods. A brief outlined summary of the components and the principles follows.

(i) Components

- Sample/weighing chamber: transparent glass enclosure with sliding door/sash where filter is placed and shielded from moving airstreams and related turbulence during weighing
- Weighing pan: stage on which filter is placed for weighing
- Mechanical or electrical balance mechanism: mechanical or electrical device that responds to weight of filter
- Readout: responds to mechanical or electrical signal from balance mechanism and provides weight of filter

(ii) Principle of Analysis

- Following sampling and prior to analysis, the orifice plugs are removed from the cassette that is placed in a dessicator to allow time for absorption of water that may have sorbed to the particulate and filter during sampling.
- Following dessication, the cassette is disassembled and the filter is removed and weighed to determine the post-weight to the nearest 0.001 mg.
- To minimize the attraction and contamination of the filter and sample, they are passed through an ionization device to eliminate static charge during pre- and post-sampling weighing.

(iii) Determination of Concentration of Total Dust

- Determine the amount of total particulate collected by subtracting filter pre-weight from filter post-weight and express amount as weight in milligrams (mg).

$$\text{Total Particulate (mg)} = \text{Filter Post-weight (mg)} - \text{Filter Pre-weight (mg)} \quad (4.3)$$

- Determine the sampled volume of air by multiplying sampling time (T [min]) by average sampling flow rate (Q [l/min]) and convert liters (l) to cubic meters (m^3).

$$\textit{Air Volume}\ (m^3) = T\,(\min) \times Q_{avg}\ (l/\min) \times \frac{1\ m^3}{10^3\ l} \qquad (4.4)$$

- Determine the concentration of total particulate by dividing the weight (mg) by sampled volume of air (m^3) and express concentration in units of milligrams of total particulate per cubic meter of air (mg/m^3).

$$\text{Total Particulate } (mg/m^3) = \frac{\text{Total Particulate (mg)}}{\text{Air Volume } (m^3)} \qquad (4.5)$$

- Note: The adjusted concentration (C) can be calculated by subtracting the weight of particulate measured on a blank PVC filter (B) from the sample PVC filter (S); dividing by air volume sampled (V); and correcting for sampling efficiency (E), which is equal to ≤1 depending if 100% or less efficiency is attained.

$$C = \frac{S - B}{E\,V} \qquad (4.6)$$

UNIT 4 EXERCISE

OVERVIEW

The exercise will provide the fundamental concepts for conducting integrated sampling for total particulate using a filter medium. In addition, a common and related analytical method using an electrobalance is introduced. Thus, the exercise will focus on sampling and analysis of total particulate. The material and methods are based on a modification of National Institute for Occupational Safety and Health (NIOSH) Method 0500. Place a check mark (✔) in the open box (☐) when you have obtained applicable material and completed the steps for sampling and analytical methods.

MATERIAL

1. Calibration of Air Sampling Pump

☐ Multi- or high-flow air sampling pump
☐ 37-mm, 5.0-µm pore size PVC filter with cellulose support pad in assembled three-stage plastic cassette
☐ Two 2- to 3-ft lengths of 1/4-in. i.d. flexible hose
☐ Manual or electronic frictionless bubble-tube calibrator
☐ Soap solution (depending on calibrator used)
☐ Stopwatch (if manual calibrator is used)
☐ Calibration data form (Figure 4.6)

2. Sampling for Total Particulate

☐ Antistatic device and electrobalance to prepare (pre-weigh) filter for sample collection
☐ Pre-weighed 37-mm, 5.0-µm pore size PVC filter with cellulose support pad in assembled three-stage plastic cassette

Calibration Data Form:
Low-Flow, High-Flow, and Multi-Flow Air Sampling Pumps

Name and Location of Calibration: ____________________

Calibration Conducted By: ____________________ Date Calibration Conducted: ________

Calibration Instruments (Type/Manufacturer/Model): ____________________

Air Sampling Pump
(Type/Manufacturer/Model): ____________________

Collection Medium In-Line: ____________ Calibrator Volume: ________cc

Air Temperature: ____°C Air Pressure: ____mm Hg Relative Humidity: ____%

Pump No.	Pre-Sampling Calibration Flow Rate (Q_{pre})					Post-Sampling Calibration Flow Rate (Q_{post})					Average Flow Rate
	Time (sec)				Q_{pre}	Time (sec)				Q_{post}	Q_{avg}
	T_1	T_2	T_3	T_{avg}	L/min	T_1	T_2	T_3	T_{avg}	L/min	L/min

Calibration Notes:

Figure 4.6 Calibration data form for high-flow air sampling pump.

- ☐ Orifice plugs
- ☐ Cellulose cassette bands or tape to secure stages of the cassette (optional)
- ☐ Multi-flow or high-flow air sampling pump
- ☐ 1 3-ft length of 1/4-in. i.d. flexible hose
- ☐ Cassette adaptors to connect flexible hose to orifice of cassette
- ☐ Field sampling data form (Figure 4.7)

3. Analysis of Sample for Total Particulate

- ☐ Electrobalance accurate to ≥0.001 mg
- ☐ Laboratory analysis data form (Figure 4.8)

METHOD

1. Pre-sampling Calibration of Air Sampling Pump

- ☐ Remove the pump from charger, turn "ON," and allow it to operate for 5 min prior to calibrating.
- ☐ Complete a calibration form, remembering to record name, location, date, air temperature, air pressure, pump manufacturer and model, calibration apparatus, times for three trials, and average flow rate for pre- and post-sampling calibrations.
- ☐ Obtain a representative sampling medium for use during calibration of the air sampling pump; this should consist of a 37-mm, 5.0-μm pore size PVC filter (unweighed or weighed) and cellulose support pad in assembled three-stage plastic cassette.
- ☐ Insert a cassette adapter into one end of a 2- to 3-ft length of flexible hose.
- ☐ Connect the other end of the flexible hose to the air sampling pump.
- ☐ Connect the adapter end of the flexible hose to the outlet orifice located on the first stage of the support pad side of the cassette used as the representative sampling medium.
- ☐ Insert another cassette adapter into one end of another 2- to 3-ft length of flexible hose.
- ☐ Connect the adapter end of the second flexible hose to the inlet orifice located on the third stage or filter side of the cassette used as the representative sampling medium.
- ☐ Connect the remaining open end of the flexible hose to a manual or electronic frictionless bubble-tube calibration device; this constitutes the calibration train.
- ☐ Aspirate the bubble solution into the calibration device to form a soap film (bubble).
- ☐ Record the time (T [sec]) it takes the soap film (bubble) to traverse a volume of 500 to 1000 ml.
- ☐ If a manual bubble-tube is used, calculate the flow rate (Q [l/min]). Otherwise, read the flow rate directly from the electronic bubble-tube calibrator.
- ☐ If the flow rate is not 2 l/min ±5%, adjust the flow control on the pump and repeat calibration procedure until desired flow rate is achieved.
- ☐ Once the desired flow rate is achieved, in this example 2 l/min ±5%, repeat the calibration check two to three times and calculate the average pre-sampling flow rate.
- ☐ Disconnect the pump from the remainder of the calibration apparatus and turn "OFF." The calibrated air sampling pump is ready for use.

2. Prepare Sampling Media

- ☐ Complete a laboratory analysis data form, remembering to record name, location, date, air temperature, air pressure, analyte (e.g., total particulate), electrobalance manufacturer and model, filter type, filter identification, and filter pre-weight.
- ☐ Zero the electrobalance.
- ☐ Check the calibration of the electrobalance using a standardized weight; if okay, then proceed.
- ☐ Remove the filter package from dessicator.
- ☐ Using a forceps, gently remove a PVC filter from the package and pass it across the ionization device prior to placing it on the weighing pan.

Field Monitoring Data Form:
Integrated Monitoring for Total and Respirable Particulates (and Silica)

Facility Name and Location: ______________________ **Contaminant Sampled:** ☐ **Total Particulate**
☐ **Respirable Particulate**
☐ **Crystalline Silica (form)** ____________

Monitoring Conducted By: ______________________

Collection Medium: ______________

Date Monitoring Conducted: ____________

Method(s)(Source/Number/Name): ______________________

Monitoring Instruments (Type/Manufacturer/Model): ______________________

Air Temperature: ____°C **Air Pressure:** ____mm Hg **Relative Humidity:** ____%

Identification of Personnel or Area	Field Sample	Pump No.	Pre-Sample Calibration Flow Rate	Post-Sample Calibration Flow Rate	Average Flow Rate (Q)	Sample Start-Time	Sample Stop-Time	Total Sample Time (T)	Volume Sampled (Vol)

Field Notes:

Figure 4.7 Field sampling data form for total particulate.

Laboratory Analysis Data Form:
Electrobalance Gravimetric Analysis of Total Particulates (and Percent Silica)

Monitored Facility Name and Location: ______________________ **Contaminant Sampled:** ______________

Monitoring Conducted By: ______________________ **Date Monitoring Conducted:** ______________

Laboratory Facility Name and Location: ______________________ **Contaminant(s) Analyzed:**
☐ **Total Particulate**

Method (Source/Number/Name): ______________________ ☐ **Crystalline Silica (form)** ______________

Analysis Conducted By: ______________________ **Date Analysis Conducted:** ______________

Filter Collection Medium (Type/Manufacturer/Lot No.): ______________________

Electrobalance (Type/Manufacturer/Model): ______________________

X-Ray Diffraction (Type/Manufacturer/Model): ______________________

Air Temperature: ____°C **Air Pressure:** ____mm Hg **Relative Humidity:** ____%

Laboratory Sample Identification No.	Field Sample Identification No.	Pre-Sampling Filter Weight (mg)	Post-Sampling Filter Weight (mg)	Post-Sample Wt. Minus Pre-Sample Wt. (mg)	Average Flow Rate (Q) (L/min)	Total Sample Time (T) (min)	Volume Sampled (Vol) (m^3)	Percent Silica (form)	Conc. Particulate (mg/m^3)	
									Total	Silica

Laboratory Notes:

Figure 4.8 Laboratory analysis data form for gravimetric analysis of total particulate.

- ☐ Read the weight from the electrobalance to the nearest 0.001 mg.
- ☐ Weigh the filter a second time. If the first and second weighings differ by >0.005 mg, repeat zeroing and calibration of the electrobalance to assure that it is functioning properly.
- ☐ Assuming that the two weighings do not differ by >0.005 mg, record the average as "filter pre-weight" to the nearest 0.001 mg.
- ☐ Using a forceps, place an unweighed cellulose support pad into the outlet stage of the three-stage cassette.
- ☐ Using a forceps, gently remove the pre-weighed filter from the weighing pan of the electrobalance and place it on the support pad.
- ☐ Place the center ring of the three-stage cassette on the filter.
- ☐ Place the inlet stage onto the center stage.
- ☐ Press the three stages of the cassette together by using your palm and a flat rigid plastic or metal plate to evenly apply pressure.
- ☐ Place the orifice plugs into the ports in the inlet and outlet stages of the cassette.
- ☐ Place a wet cellulose band around the periphery of the cassette and allow to dry and shrink; tape can be used instead (optional step).
- ☐ Label the cassette with a sample identification number.

3. Sampling for Airborne Total Particulate

- ☐ Get approval to conduct sampling in an area where airborne levels of particulate are likely to be detectable and measurable, and perform sampling accordingly. Alternatively, conduct a simulated sampling exercise and obtain a sample spiked with particulate from your instructor for subsequent analysis.
- ☐ Complete a field sampling data form, remembering to record name, location, person or area sampled, date, air temperature, air pressure, relative humidity, analyte (e.g., total particulate), air sampling pump manufacturer and model, pump identification, sampling medium, sample identification, flow rate, start and stop times for sampling, sampling duration, and air volume sampled.
- ☐ Turn "ON" pump and let operate for 5 min prior to connecting flexible hose and sample collection medium.
- ☐ Insert a cassette adapter into the end of flexible hose that will be attached to pump.
- ☐ Place and secure pump in area where the sample will be collected.
- ☐ Remove orifice plugs from a cassette and attach outlet port of cassette to adaptor positioned in flexible hose that will be connected to pump.
- ☐ Connect flexible hose to pump and secure sample collection medium.
- ☐ Record the sample identification number and start time and allow sample to run for at least 2 h depending on conditions.
- ☐ After an acceptable sampling period has elapsed, turn "OFF" pump, record stop time, and calculate time (T [min]).
- ☐ Remove cassette from the sampling train and insert the orifice plugs in the inlet and outlet stages. Transport sample to laboratory.
- ☐ Carefully wipe the outside of the cassette to remove any surface particulate that may have adhered to the surface.
- ☐ Remove orifice plugs and place sample in dessicator for approximately 24 h prior to analysis.

4. Post-sampling Calibration of Air Sampling Pump

- ☐ Repeat steps summarized in Method #1, except do not adjust the flow rate of the pump.
- ☐ The post-sampling flow rate (Q_{post}) should be within ±5% of the pre-sampling flow rate (Q_{pre}).
- ☐ Determine the average sampling flow rate (Q_{avg}) by averaging the pre- and post-sampling flow rates, and express average flow rate as liters per minute.

5. Analysis of Sample for Total Particulate

- ☐ Zero the electrobalance and check calibration using a standardized weight.
- ☐ Remove the sample from dessicator and remove the cellulose band or tape from around the periphery of the cassette if present.
- ☐ Disassemble the cassette by carefully removing the middle and inlet stages to expose the filter.
- ☐ Using a probe, gently push through the outlet stage orifice so that the support pad and filter are elevated and accessible for grasping with a forceps.
- ☐ Using a forceps, gently remove the filter and place it on the weighing pan of the electrobalance.
- ☐ Record the weight as filter post-weight to the nearest 0.001 mg.
- ☐ Calculate concentration of total particulate units of milligrams per cubic meter.
- ☐ Complete a laboratory analysis data form, remembering name, location, date, air temperature, air pressure, analytes (total particulate), elctrobalance manufacturer and model, sample identification, filter pre-weight, filter post-weight, total particulate weight, air volume sampled, and concentration.

UNIT 4 EXAMPLES

Example 4.1

A high-flow air sampling pump was calibrated using a manual frictionless bubble-tube. What was the flow rate (Q) in liters per minute (l/min) if it took an average (n = 3 trials) of 15 sec (T) for the bubble to traverse a volume (V) of 500 cm^3?

Solution 4.1

$$Q\ (\mathrm{l/min}) = \frac{\mathrm{V\ (cm^3)}}{\mathrm{T\ (sec)}} \times \frac{60\ \mathrm{sec}}{\mathrm{min}} \times \frac{1\ \mathrm{l}}{10^3\ \mathrm{cm^3}}$$

$$= \frac{500\ \mathrm{cm^3}}{15\ \mathrm{sec}} \times \frac{60\ \mathrm{sec}}{\mathrm{min}} \times \frac{1\ \mathrm{l}}{10^3\ \mathrm{cm^3}}$$

$$= 2\ \mathrm{l/min}$$

Example 4.2

A sample was collected for 2.5 h (T [min]) using a high-flow sampling pump at a an average flow rate (Q) of 2 l/min. What was the volume (V [m^3]) of air sampled by the pump (Note: 2.5 h = 150 min)?

Solution 4.2

$$\mathrm{V\ (m^3)} = Q\ (\mathrm{l/min}) \times T\ (\mathrm{min}) \times \frac{1\ \mathrm{m^3}}{10^3 l}$$

$$= 2\ (\mathrm{l/min}) \times 150\ \mathrm{min} \times \frac{1\ \mathrm{m^3}}{10^3 l}$$

$$= 0.300\ \mathrm{m^3}$$

Example 4.3

A 0.300 m^3 volume of air was sampled for total dust using a high-flow air sampling pump connected to a three-stage plastic cassette with a pre-weighed 37-mm, 5.0-µm PVC filter and support and medium. Laboratory analysis of the filter using gravimetry indicated that the filter pre-sampling weight was 14.616 mg and post-sampling weight was 16.517 mg. What was the concentration (mg/m^3) of total dust in the air during the sampling period?

Solution 4.3

$$\text{Total Dust (mg / m}^3) = \frac{\text{Post - weight (mg)} - \text{Pre - weight (mg)}}{\text{Air Volume (m}^3)}$$

$$= \frac{16.517 \text{ mg} - 14.616 \text{ mg}}{0.300 \text{ m}^3}$$

$$= \frac{1.901 \text{ mg}}{0.300 \text{ m}^3}$$

$$= 6.34 \text{ mg / m}^3$$

UNIT 4 CASE STUDY: SAMPLING AND ANALYSIS OF AIRBORNE TOTAL PARTICULATE

OBJECTIVES

Upon completion of Unit 4, including sufficient reading and studying of this and related reference material, and review of the short case study and hypothetical industrial hygiene sampling and analytical data presented below, learners will be able to:

- Calculate the concentrations and time-weighted averages (TWAs).
- Interpret the data and results.
- Concisely summarize the information in a concise report.
- Retain the general scientific and technical principles and concepts related to the topic.

INSTRUCTIONS

After reading the short case study, locate and read the applicable NIOSH method for sampling and analysis (http://www.cdc.gov/niosh/nmam/nmammenu.html). Next, review the data shown in the tables below and perform the necessary calculations to complete the tables (assume normal temperature pressure). Review and interpret the results and compare calculated TWAs to applicable occupational exposure limits (e.g., OSHA-PEL, ACGIH-TLV). Also, prepare responses to the questions that follow the data section. Finally, prepare a fictional, but representative, report following the outline or similar format shown in Appendix B.

CASE STUDY

A group of workers at a plastic product manufacturing facility are responsible for manually emptying 50-lb. bags of powdered polyvinyl chloride (PVC) resin into large mixing reactors. Most of the PVC dust generated appears to be captured by the local exhaust slot hoods mounted adjacent to the reactor openings. There is still some observable airborne dust generated during the bag emptying process. In addition, after emptying each bag, workers are observed throwing the bags onto flatbed carts for eventual transport to an on-site incinerator. Although considered "empty," the bags contain some residual powdered PVC resin. Thus, additional airborne dust is generated when the empty bags are thrown onto the carts and when workers manually push down on the empty bags to compress the stack. Accordingly, a certified industrial hygienist (CIH) deemed it necessary to conduct integrated personnel and area air sampling for PVC dust as total dust. Following sampling, he prepared samples and field blanks prior to delivering them to an American Industrial Hygiene Association (AIHA) accredited laboratory for analysis. Sampling and analysis for total dust followed NIOSH Method 0500. Calibration, field, and laboratory data are summarized in the tables below:

DATA AND CALCULATIONS

Table 4.1 Calibration and Field Data

Name of Personnel or Area	Sample No.	Analyte	Avg. Pre-sample Q (l/min)	Avg. Post-sample Q (l/min)	Avg. Q (l/min)	Sample Start Time	Sample Stop Time	Sample Time (min)	Air Volume Sampled (l)
Jim K.	2155	Total dust	Pump No. 3021: 2.00	Pump No. 3021: 2.00		0700	1059		
	2157					1100	1500		
Mixer 1	2136	Total dust	Pump No. 3023: 2.01	Pump No. 3023: 2.04		0701	1101		
	2135					1103	1459		
Blank	2130	Total dust	N/A	N/A					

N/A = Not applicable.

Table 4.2 Lab Data

Name of Personnel or Area	Sample No.	Analyte	Pre-sample Weight (mg)	Post-sample Weight (mg)	Weight Total Dust (mg)	Percent Crystalline Quartz (%)	Air Volume Sampled (m^3)	Conc. Total Dust (mg/m^3)
Jim K.	2155	Total dust	11.016	15.643		N/A		
	2157		11.000	15.444		N/A		
Mixer 1	2136	Total dust	11.101	13.756		N/A		
	2135		11.099	14.041		N/A		
Blank	2130	Total dust	11.011	11.011	<LOD	N/A		

LOD = Limit of detection.
N/A = Not applicable.

Table 4.3 Calculated Time-Weighted Averages

Name of Personnel or Area	Sample No.	Analyte	8-hr TWA Total Dust (mg/m^3)
Jim K.	2155	Total dust	
	2157	Total dust	
Mixer 1	2136		
	2135	Total dust	
Blank	2130		

QUESTIONS

- Were the average pre- and post-sampling flow rates within ±5% of each other?
- Was the total sample volume collected within the min-max range specified in the applicable NIOSH method?
- What is the minimum number of field blanks that should be submitted to the laboratory, based on the number of samples collected and the applicable NIOSH method?
- What considerations or assumptions did you make when calculating the TWAs? For example, did you assume the workers took breaks (e.g., lunch) in areas outside the immediate area? Would you continue monitoring while workers are on breaks, even if outside the potential exposure area? If not, what if the actual sample collection period were less than 8 h (480 min)?
- What is the estimated limit of detection (LOD) for the analyte as per the applicable NIOSH method?
- How did the calculated TWAs compare to applicable occupational exposure limits, such as the OSHA-PEL and ACGIH-TLV?
- What was the basis for the established occupational exposure limits? (Learners and practitioners alike are strongly encouraged to obtain and read applicable information found in resources such as the NIOSH *Criteria Documents* and ACGIH *Documentation of the TLVs and BEIs*.)
- Are there other NIOSH or OSHA air sampling and analytical methods for the same agent? If so, what are the reference numbers and what are the differences relative to sample collection (e.g., medium used) and analysis (e.g., analytical instrument used)?

REPORT

Prepare a concise report (see Appendix B for outline of sample format) that includes typed text and tables (and figures if you desire). Your report needs to include a section on calibration (how you calibrate pump with representative media), sample collection (instruments, media, and NIOSH method used), and analysis (instrument and NIOSH method used). In other words, write your short field report so the reader knows what was done to prepare for sampling and analysis for respirable particulate (including crystalline silica). Also, summarize, interpret, and discuss the results; state your conclusions; and make applicable recommendations. Remember to add a bibliography at the end of the report citing any applicable book, journal, and Internet references that you consult. Finally, append a copy of the applicable NIOSH methods.

UNIT 5

Evaluation of Airborne Respirable Particulate: Integrated Personal and Area Monitoring Using an Air Sampling Pump with a Polyvinyl Chloride Filter Medium and Dorr-Oliver Cyclone

LEARNING OBJECTIVES

At the completion of Unit 5, including sufficient reading and studying of this and related reference material, learners will be able to correctly:

- Name, identify, and assemble the components of a common sampling train, including the specific sampling medium, used for conducting integrated sampling of airborne respirable nuisance dust or particulates not otherwise classified, hereafter referred to as respirable particulate.
- Name, identify, and assemble the applicable calibration train that uses a primary standard for measuring flow rate of a multi- or high-flow air sampling pump.
- Calibrate a multi- or high-flow air sampling pump, with the applicable representative sampling medium in-line, using a manual or electronic frictionless bubble-tube.
- Summarize the principles of sample collection of airborne respirable particulate.
- Conduct integrated sampling of airborne respirable particulate.
- Prepare samples for analysis of collected particulate using gravimetry.
- Name and identify the components of a mechanical or electrical balance.
- Name, identify, and assemble the components necessary for analysis of samples for respirable particulate using an electrobalance.
- Check the calibration of an electrobalance using standard weights.
- Summarize the principles of sample analysis of filters used for collection of airborne respirable particulate.
- Conduct analysis of a sample for respirable particulate using an electrobalance.
- Perform applicable calculations and conversions related to pre- and post-sampling flow rate of an air sampling pump, average flow rate of an air sampling pump, pre- and post-weight of a filter medium, sampling time, sampled air volume, weight of respirable particulate collected, and concentration of respirable particulate in units of milligrams per cubic meter (mg/m^3).
- Record all applicable calibration, sampling, and analytical data using calibration, field sampling, and laboratory analysis data forms.

OVERVIEW

There are situations in which knowing the concentration of particulate with the potential to reach the gas-exchange (i.e., alveolar) regions of the lungs is important to adequately characterize the hazard associated with exposure to airborne particulates. The respirable mass fraction refers to the particulates with diameters <10 μm. Occupational exposure limits for respirable particulate are two to three times lower than limits established for total particulate (i.e., total dust, particulates not otherwise specified).

A filter medium in combination with a particle size-selective device, such as the Dorr-Oliver Cyclone, are used to separate and collect the respirable fraction from inhalable or total particulates. The particulate that collects on the filter medium represents the respirable fraction of airborne particulate. It is important to note that the respirable fraction also can be a specific element or compound or a mixture of both highly toxic and practically non-toxic components. Crystalline silica (e.g., quartz) is an example of a relatively highly toxic dust that is commonly sampled relative to the respirable fraction using a size-selective device (e.g., cyclone).

FILTER MEDIUM AND PARTICLE SIZE-SELECTIVE DEVICE FOR RESPIRABLE PARTICULATE

The sample medium for collecting respirable particulate is the same as that used to collect total particulate — a 37-mm diameter polyvinyl chloride (PVC) filter with 5.0-μm pore size. In addition, the principle of deposition and collection of particulate on the filter is the same. The requirement for desiccating the filters is also the same. The filter medium for collecting respirable particulate is positioned on a cellulose support pad and often secured within only a two-stage (Figure 5.1) instead of a three-stage 37-mm diameter plastic cassette. The bottom stage contains the support pad and the filter. It contains an orifice (hole) in the center that serves as the outlet port for sampled air to exit the cassette. A center stage ring typically is not used. The inlet stage contains a hole in the center that serves as the inlet port for sampled air to enter the cassette.

The main difference when sampling for the respirable fraction is that theoretically only particle sizes <10 μm flow into the cassette and contact the filter. The two-stage cassette with filter medium is positioned within a particle size-selective device. A common size-selective device, known as a Dorr-Oliver Cyclone (Figure 5.2), directs particles with diameters <10 μm into the inlet orifice of

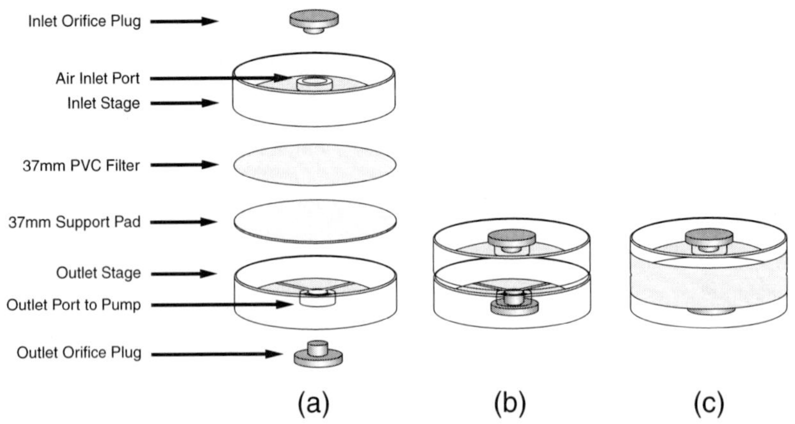

Figure 5.1 A 37-mm two-stage cassette with a 37-mm PVC filter and support pad. (a) Exploded view, (b) assembled view, and (c) assembled view secured with shrink band or tape.

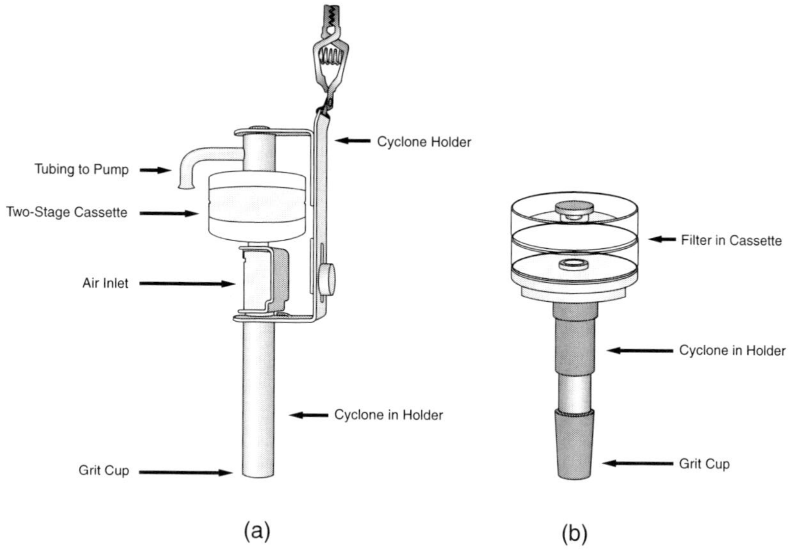

Figure 5.2 (a) Dorr-Oliver Cyclone assembly with a two-stage 37-mm cassette containing a 37-mm PVC filter on support pad. (b) Example of aluminum cyclone assembly with a two-stage 37-mm cassette containing a 37-mm PVC filter on support pad.

the cassette. Air enters the cyclone creating a downward vortex, and due to centrifugal forces, the larger and denser particles flow downwardly and settle on the interior sides or travel into the bottom of the device called a grit cup. Simultaneously and optimally, an upward vortex transports respirable particles with smaller diameters (<10 μm) in the opposite direction. The respirable particles enter the inlet port of the positioned two-stage cassette and deposit and collect on a pre-weighed PVC filter.

Various particle size-selective devices are available, differing mainly based on the designs and related sampling cut-points and flow rates[1,2]. The cut-point refers to the particle size or diameter of the particles that cyclones and other size-selective devices collect with 50% sampling efficiency. The Dorr-Oliver Cyclone has a 50% cut-point of 4.0 μm. Collection efficiencies are higher and lower for particle sizes less than or greater than, respectively, the cut-point. Common flow rates and corresponding cut-points for other relatively common cyclones are summarized in Table 5.1. Theoretically, size-selective devices with a higher cut-point will collect more particulate. For example, a cyclone with a cut-point of 4.0 μm will collect more particulate mass than a cyclone with a cut-point of 3.5 μm.

Accordingly, size-selective devices for collecting the inhalable particulate mass fraction, discussed briefly in Unit 4, would have an even higher cut-point. For example, the IOM sampler is a size-selective device for inhalable particulate and has a cut-point of a 100 μm to collect a wider range (i.e., fine and coarse) of particle sizes.

Table 5.1 Types of Size-Selective Cyclones and Sampling Parameters

Cyclone Type	Specific Flow Rate (l/min)	50% Cut-Point (μm)
Dorr-Oliver	1.7	4.0
Higgins-Dewell	2.2	5.0
Aluminum	2.5	5.0

[1] Vincent, J.H., *Particle Size-Selective Sampling for Particulate Air Contaminants,* The American Conference of Governmental Industrial Hygienists, Cincinnati, OH, 1999.
[2] Johnson, D.L. and Swift, D.L., Sampling and sizing particles, in *The Occupational Environment — Its Evaluation and Control*, DiNardi, S., Ed., AIHA Press, Fairfax, VA, 1997.

SAMPLING

The method for measuring respirable particulate is somewhat similar to the method used to measure total particulate in the air. The most significant difference, however, is the method of sampling since a size-selective device is needed for separating respirable particulate. Gravimetric analysis is used to measure respirable particulate, and the procedure is virtually the same as that described for total particulate.

The PVC filter is pre-weighed using an electrobalance and positioned on a cellulose backup pad in a two-stage plastic cassette. The cassette is assembled and capped on the inlet and outlet ports. When used for sampling, the cassette is uncapped and positioned in a nylon Dorr-Oliver Cyclone apparatus attached to an air sampling pump using a length of $^1/_4$ in. internal diameter (i.d.) flexible hose (Figure 5.3).

A typical sampling rate is 1.7 l/min ±5%. Sampling flow rates must be accurate when using the cyclone since the device is designed to collect and separate the respirable from the non-respirable mass fraction most efficiently at the prescribed flow rate. If the flow rate is >1.7 l/min, large particles will not be separated by the cyclone and will be collected on the filter yielding a false high. Alternatively, if flow rates are less than 1.7 l/min, this will cause some of the respirable dust particles to be removed by the cyclone, and results will show lower concentrations of respirable dust than what actually exists.

A brief outlined summary of sample collection follows.

(i) Calibration

- An air sampling pump is calibrated using a manual or electrical calibrator with representative media (i.e., an assembled two-stage cassette containing a 37-mm, 5.0-µm pore size PVC filter and support pad connected to a clean nylon Dorr-Oliver Cyclone) in-line and positioned in a sealed 2-l calibration jar (Figure 5.4).

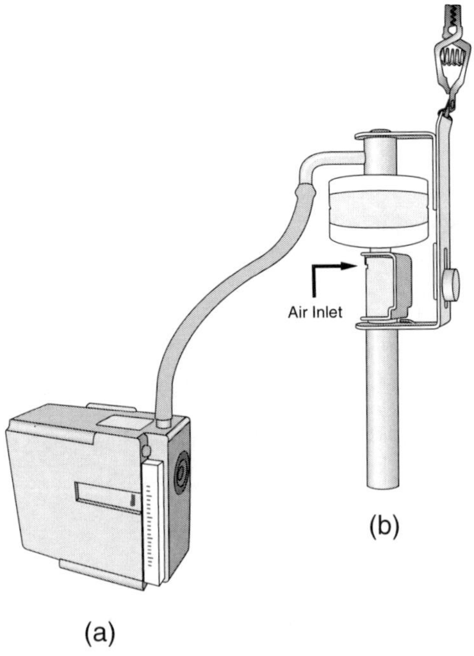

Figure 5.3 Sampling train for respirable particulate composed of (a) a high-flow pump connected with flexible hose to (b) a 37-mm PVC filter in two-stage cassette positioned within a Dorr-Oliver Cyclone assembly.

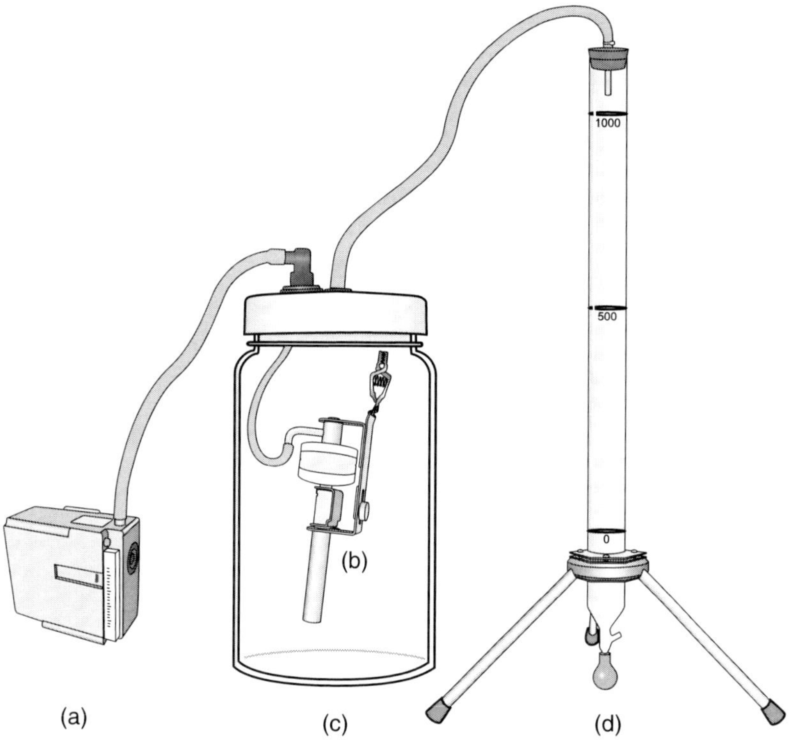

Figure 5.4 Calibration train for respirable particulate composed of (a) a high-flow pump connected with flexible hose to (b) a 37-mm PVC filter in three-stage cassette positioned within a Dorr-Oliver Cyclone assembly contained within (c) a calibration jar in-line with (d) a frictionless bubble-tube.

- Pre- (Q_{pre}) and post-sampling (Q_{post}) flow rates are determined by measuring the average time (T_{avg} sec) based on three trials ($\Sigma\ T_{1,2,3}/3$) for a bubble to traverse a specific volume (V [cm^3]) of the bubble-tube.

$$Q\ (l/\min) = \frac{V\ (cm^3)}{T_{avg}(sec)} \times \frac{60\ sec}{1\min} \times \frac{1\ l}{10^3 cm^3} \tag{5.1}$$

- The average flow rate (Q_{avg}) is based on the average of pre- (Q_{pre}) and post-sampling (Q_{post}) flow rates.

$$Q_{avg} = \frac{Q_{pre} + Q_{post}}{2} \tag{5.2}$$

- The flow rate for sampling respirable particulate is 1.7 l/min.

(ii) Preparation for Sampling

- Remove the Dorr-Oliver Cyclone from the stainless steel holder and dissamble it into its separate components. Examine and assure that the cyclone and its grit cap are clean and in good condition. Indeed, before and after use, clean the cyclone in warm soapy water. Alternatively, immerse and wash in an ultrasonic bath. Do not insert devices (e.g., pipe cleaners, small brushes) into the cyclone during cleaning since they can scratch and damage the lumen of the vortex. Rinse with fresh water and let air dry. Carefully reassemble the cyclone and insert into its holder.

- Weighed PVC filter plus unweighed support pad are inserted into a plastic, two-stage cassette which is then assembled, labeled, and connected to the Dorr-Oliver Cyclone.
- The sampling train is assembled and consists of multi- or high-flow pump; $^1/_4$ in. internal diameter (i.d.) flexible hose with clips; and a 37-mm two-stage plastic cassette containing a weighed 37-mm, 5.0-μm pore size filter on an unweighed cellulose support pad connected to a nylon Dorr-Oliver Cyclone.

(iii) Conducting Sampling

- A labeled plastic cassette is positioned with the inlet port facing downward in the cyclone assembly, which is attached within the breathing zone, and the air sampling pump is clipped to the belt of the worker for personal sampling. The sampling train is positioned and secured in specific location for area sampling.
- The pump is turned "ON" and start time is recorded.
- After specified time, the pump is turned "OFF" and stop time is recorded.
- Remove the plastic cassette from the cyclone assembly and insert orifice plugs into the inlet and outlet ports. Transport the sample to the laboratory for analysis.

ANALYSIS

The analytical method for respirable particulate is identical to the method for total particulate. Analysis of the sample involves a gravimetric method using a mechanical or electrical balance, as described for total particulate in Unit 4. The gravimetric procedure involves comparison of post- and pre-sampling weight of filter. Pre-weight is theoretically the weight of the filter, and post-weight is the weight of the filter plus the collected particulate with diameters <10 μm. The difference in weights (post-minus pre-weight) is the weight of respirable particulate collected. Gravimetric analysis is the most simplistic of the major analytical methods. A brief outlined summary of the components appears in Unit 4. The principles for analyzing samples for respirable particulate follows.

(i) Principle of Analysis

- Following sampling and prior to analysis, the orifice plugs are removed from the cassette that is then placed in a dessicator to allow time for absorption of water that may have sorbed to the particulate and filter during sampling.
- Following dessication, the cassette is disassembled and the filter is removed and weighed to determine the post-weight to the nearest 0.001 mg.
- To minimize the attraction and contamination of the filter and sample, they are passed through an ionization device to eliminate static charge during pre- and post-sample weighing.

(ii) Determination of Concentration

- Determine the amount of respirable particulate collected by subtracting filter pre-weight from filter post-weight and express the amount as weight in milligrams (mg).

$$\text{Respirable Particulate (mg)} = \text{Filter Post-weight (mg)} - \text{Filter Pre-weight (mg)} \qquad (5.3)$$

- Determine the sampled volume of air by multiplying sampling time (T [min]) by average sampling flow rate (Q [l/min]) and convert liters (l) to cubic meters (m^3).

$$\text{Air Volume } (m^3) = T\,(\text{min}) \times Q_{avg}\,(l/\text{min}) \times \frac{1\,m^3}{10^3\,l} \tag{5.4}$$

- Determine the concentration of respirable particulate by dividing the weight (mg) by sampled volume of air (m^3). Express concentration in units of milligrams of respirable particulate per cubic meter of air (mg/m^3).

$$\text{Respirable Particulate } (\text{mg}/\text{m}^3) = \frac{\text{Respirable Particulate (mg)}}{\text{Air Volume } (\text{m}^3)} \tag{5.5}$$

- Note: The adjusted concentration (C) can be calculated by subtracting the weight of particulate measured on a blank PVC filter (B) from the sample PVC filter (S); dividing by air volume sampled (V); and also correcting for sampling efficiency (E), which is equal to ≤ 1 depending if 100% or less efficiency is attained.

$$C = \frac{S - B}{EV} \tag{5.6}$$

UNIT 5 EXERCISE

OVERVIEW

The exercise will provide the fundamental concepts for conducting integrated sampling for respirable particulate using a filter medium and respirable dust cyclone. In addition, a common and related analytical method using an electrobalance is introduced. Thus, the exercise will focus on sampling and analysis or respirable particulate. The material and methods are based on a modification of National Institute for Occupational Safety and Health (NIOSH) Method #0600. The method is applicable for use with the Dorr-Oliver, Higgins-Dewell (H-D), or aluminum cyclones. Place a check mark (✔) in the open box (☐) when you have obtained applicable material and completed the steps for sampling and analytical methods.

MATERIAL

1. Calibration of Air Sampling Pump

☐ Multi- or high-flow air sampling pump
☐ 37-mm, 5.0-µm pore size PVC filter with cellulose support pad in a two-stage plastic cassette positioned within a nylon cyclone assembly
☐ Two 2- to 3-ft lengths of 1/4 in. i.d. flexible hose
☐ 2-l calibration jar
☐ Manual or electronic frictionless bubble-tube calibrator
☐ Soap solution
☐ Stopwatch (if manual calibrator is used)
☐ Calibration data form (Figure 5.5)

2. Sampling for Respirable Particulate

☐ Antistatic device and electrobalance to prepare (pre-weigh) filter for sample collection

Calibration Data Form:
Low-Flow, High-Flow, and Multi-Flow Air Sampling Pumps

Name and Location of Calibration: ____________________

Calibration Conducted By: ____________________ **Date Calibration Conducted:** ________

Calibration Instruments (Type/Manufacturer/Model): ____________________

Air Sampling Pump (Type/Manufacturer/Model): ____________________

Collection Medium In-Line: ____________ **Calibrator Volume:** ________cc

Air Temperature: ____°C **Air Pressure:** ____mm Hg **Relative Humidity:** ____%

Pump No.	Pre-Sampling Calibration Flow Rate (Q_{pre})					Post-Sampling Calibration Flow Rate (Q_{post})					Average Flow Rate
	Time (sec)				Q_{pre}	Time (sec)				Q_{post}	Q_{avg}
	T_1	T_2	T_3	T_{avg}	L/min	T_1	T_2	T_3	T_{avg}	L/min	L/min

Calibration Notes:

Figure 5.5 Calibration data form for high-flow air sampling pump.

- ☐ Preweighed 37-mm, 5.0-μm pore size PVC filter with cellulose support pad in a two-stage plastic cassette positioned within a nylon Dorr-Oliver Cyclone assembly (Note: A different cyclone, such as the H-D or aluminum cyclone, can be substituted for the Dorr-Oliver Cyclone in this exercise provided the prescribed flow rate for the specific cyclone type is used.)
- ☐ Orifice plugs
- ☐ Cellulose cassette bands or tape to secure the stages of the cassette (optional)
- ☐ Multi- or high-flow air sampling pump
- ☐ One 2- to 3-ft length of $^1/_4$-in. i.d. flexible hose
- ☐ Field sampling data form (Figure 5.6)

3. Analysis of Sample for Respirable Particulate

- ☐ Electrobalance sensitive to ≤0.001 mg
- ☐ Laboratory analysis data form (Figure 5.7)

METHOD

1. Pre-sampling Calibration of Air Sampling Pump

- ☐ Remove the pump from charger, turn "ON," and allow it to operate for 5 min prior to calibrating.
- ☐ Complete a calibration data form, remembering to record name, location, date, air temperature, air pressure, pump manufacturer and model, calibration apparatus, times for three trials, and average flow rate for pre- and post-sampling calibrations.
- ☐ Obtain a representative sampling medium for use during calibration of the air sampling pump. This should consist of a 37-mm, 5.0-μm pore size PVC filter (unweighed or weighed) and cellulose support pad in a two-stage plastic cassette positioned within a Dorr-Oliver Cyclone assembly.
- ☐ Connect the flexible hose from inside the calibration jar to the arm of the cyclone apparatus.
- ☐ Place the cyclone apparatus in the calibration jar and tighten the lid securely. Wrap plastic tape around the perimeter of the sealed lid to reduce leakage of outside air.
- ☐ Connect the end of a flexible hose to the air sampling pump.
- ☐ Connect the other end of the same flexible hose to the tube projecting out of the lid of the calibration jar so that there is a connection between the cyclone apparatus to lid tube and lid tube to pump via flexible hoses.
- ☐ Connect the end of the second flexible hose to the second tube that projects from the lid of the calibration jar.
- ☐ Connect the remaining open end of the flexible hose to a manual or electronic frictionless bubble-tube calibration device; this constitutes the calibration train.
- ☐ Aspirate the bubble solution into the calibration device to form a soap film (bubble).
- ☐ Record the time (T [sec]) it takes the soap film (bubble) to traverse a volume of 500 to 1000 ml.
- ☐ If a manual bubble-tube is used, calculate the flow rate. Otherwise read the flow rate directly from the electronic bubble-tube calibrator.
- ☐ If the flow rate is not 1.7 l/min ±5%, adjust the flow control on the pump and repeat calibration procedure until desired flow rate is achieved.
- ☐ Once the desired flow rate is achieved (in this example, 1.7 l/min), repeat the calibration check two to three times and calculate the average pre-sampling flow rate.
- ☐ Disconnect the pump from the remainder of the calibration apparatus and turn "OFF." The calibrated air sampling pump is ready for use.

2. Prepare Sampling Media

- ☐ Complete a laboratory analysis data form, remembering to record name, location, date, air temperature, air pressure, analyte (e.g., respirable particulate), electrobalance manufacturer and model, filter type, filter identification, and filter pre-weight.
- ☐ Zero the electrobalance.

Field Monitoring Data Form:
Integrated Monitoring for Total and Respirable Particulates (and Silica)

Facility Name and Location: ______________________ **Contaminant Sampled:** ☐ **Total Particulate**
☐ **Respirable Particulate**
☐ **Crystalline Silica (form)** ______________

Monitoring Conducted By: ______________________

Collection Medium: ______________

Date Monitoring Conducted: ______________

Method(s)(Source/Number/Name): ______________________

Monitoring Instruments (Type/Manufacturer/Model): ______________________

Air Temperature: ____°C **Air Pressure:** ____mm Hg **Relative Humidity:** ____%

Identification of Personnel or Area	Field Sample	Pump No.	Pre-Sample Calibration Flow Rate	Post-Sample Calibration Flow Rate	Average Flow Rate (Q)	Sample Start-Time	Sample Stop-Time	Total Sample Time (T)	Volume Sampled (Vol)

Field Notes:

Figure 5.6 Field sampling data form for respirable particulate.

Laboratory Analysis Data Form:
Electrobalance Gravimetric Analysis of Respirable Particulates (and Percent Silica)

Monitored Facility Name and Location: ______________________ **Contaminant Sampled:** ______________

Monitoring Conducted By: ______________________ **Date Monitoring Conducted:** ______________

Laboratory Facility Name and Location: ______________________ **Contaminant(s) Analyzed:**
☐ **Respirable Particulate**

Method (Source/Number/Name): ______________________ ☐ **Crystalline Silica (form)** ______________

Analysis Conducted By: ______________________ **Date Analysis Conducted:** ______________

Filter Collection Medium (Type/Manufacturer/Lot No.): ______________________

Electrobalance (Type/Manufacturer/Model): ______________________

X-Ray Diffraction (Type/Manufacturer/Model): ______________________

Air Temperature: ____°C **Air Pressure:** ____mm Hg **Relative Humidity:** ____%

Laboratory Sample Identification No.	Field Sample Identification No.	Pre-Sampling Filter Weight (mg)	Post-Sampling Filter Weight (mg)	Post-Sample Wt. Minus Pre-Sample Wt. (mg)	Average Flow Rate (Q) (L/min)	Total Sample Time (T) (min)	Volume Sampled (Vol) (m^3)	Percent Silica (form)	Conc. Particulate (mg/m^3)	
									Respirable	Silica

Laboratory Notes:

Figure 5.7 Laboratory analysis data form for gravimetric analysis of respirable particulate.

- ☐ Check the calibration of the electrobalance using a standardized weight; if okay, then proceed.
- ☐ Remove the filter package from dessicator.
- ☐ Using a forceps, gently remove a PVC filter from the package and pass it across the ionization device prior to placing it on the weighing pan.
- ☐ Read the weight from the electrobalance to the nearest 0.001 mg.
- ☐ Weigh the filter a second time. If the first and second weighings differ by >0.005 mg, repeat zeroing and calibration of the electrobalance to assure that it is functioning properly.
- ☐ Assuming that the two weighings do not differ by >0.005 mg, record the average as filter pre-weight to the nearest 0.001 mg.
- ☐ Using a forceps, place an unweighed cellulose support pad into the outlet (first) stage of the two-stage cassette.
- ☐ Using a forceps, gently remove the pre-weighed filter from the weighing pan of the electrobalance and place it on the support pad.
- ☐ Place the inlet (second) stage of the two-stage cassette on the filter.
- ☐ Press the two stages of the cassette together by using your palm and a flat, rigid plastic or metal plate to evenly apply pressure.
- ☐ Place the orifice plugs in the inlet and outlet ports of the cassette.
- ☐ Place a wet cellulose band around the periphery of the cassette and allow to dry and shrink. Tape can be used instead (optional step).
- ☐ Label the cassette with a sample identification number.

3. Air Sampling for Respirable Particulate

- ☐ Get approval to conduct monitoring in an area where airborne levels of respirable particulate are likely to be detectable and measurable, and perform sampling accordingly. Alternatively, conduct a simulated monitoring exercise and obtain a sample spiked with respirable particulate from your instructor for subsequent analysis.
- ☐ Complete a field sampling data form, remembering to record name, location, person or area sampled, date, air temperature, air pressure, relative humidity, analyte (respirable particulate), air sampling pump manufacturer and model, pump identification, sampling medium, sample identification, flow rate, start and stop times for sampling, sampling duration, and air volume sampled.
- ☐ Turn "ON" pump and let operate for 5 min prior to collecting flexible hose and sample collection medium.
- ☐ Connect the arm of Dorr-Oliver cyclone assembly to flexible hose that will be connected to the pump.
- ☐ Remove orifice plugs from a cassette and attach inlet port of cassette to cyclone coupler that is attached to the vortex. Tightly secure the cassette to cyclone assembly.
- ☐ Place and secure pump in area where sample will be collected.
- ☐ Connect the flexible hose to pump and secure sample collection medium.
- ☐ Record the sample identification number and start time, and allow sample to run for at least 2 h depending on conditions.
- ☐ After an acceptable sampling period has elapsed, turn "OFF" pump, record stop time, and calculate sample time (T [min]).
- ☐ Remove cassette from the sampling train and insert the orifice plugs in the inlet and outlet stages. Transport sample to laboratory.
- ☐ Carefully wipe the outside of the cassette to remove any surface particulate that may have adhered to the surface.
- ☐ Remove the orifice plugs and place sample in dessicator for approximately 24 h prior to analysis.

4. Post-sampling Calibration of Air Sampling Pump

- ☐ Repeat steps summarized in Method #1, except do not adjust flow rate of pump.
- ☐ Postsampling flow rate (Q_{post}) should be within ±5% of the pre-sampling flow rate (Q_{pre}).
- ☐ Determine the average sampling flow rate (Q_{avg}) by averaging the pre- and post-sampling flow rates and express average flow rate as liters per minute.

5. Analysis of Sample

- ☐ Zero the electrobalance and check calibration using a standardized weight.
- ☐ Remove sample from dessicator and remove the cellulose band or tape from around the periphery of the cassette if present.
- ☐ Disassemble the cassette by carefully removing the inlet stage to expose the filter.
- ☐ Using a probe, gently push through the first-stage orifice so that the backup pad and filter are elevated and accessible for grasping with a forceps.
- ☐ Using a forceps, gently remove the filter and place it on the weighing pan of the electrobalance.
- ☐ Record the weight as filter post-weight to the nearest 0.001 mg.
- ☐ Calculate concentration of respirable particulate in units of milligrams per cubic milligrams.
- ☐ Complete a laboratory analysis data form, remembering to record name, location, date, air temperature, air pressure, analytes (respirable particulate), electrobalance manufacturer and model, sample identification, filter pre-weight, filter post-weight, respirable particulate weight, air volume sampled, and concentration.

UNIT 5 EXAMPLES

Example 5.1

A high-flow air sampling pump was calibrated using a manual frictionless bubble-tube. What was the flow rate (Q) in liters per minute (l/min) if it took an average (n = 3 trials) of 17.5 sec (T) for the bubble to traverse a volume (V) of 500 cm^3?

Solution 5.1

$$Q\,(\mathrm{l/min}) = \frac{V\,(cm^3)}{T\,(\mathrm{sec})} \times \frac{60\ \mathrm{sec}}{\mathrm{min}} \times \frac{1\ \mathrm{l}}{10^3\ cm^3}$$

$$= \frac{500\ cm^3}{17.5\ \mathrm{sec}} \times \frac{60\ \mathrm{sec}}{\mathrm{min}} \times \frac{1\ \mathrm{l}}{10^3\ cm^3}$$

$$= 1.7\ \mathrm{l/min}$$

Example 5.2

A sample was collected for 8 h (T [min]) using a high-flow sampling pump at an average flow rate (Q) of 1.7 l/min. What was the volume (V [m^3]) of air sampled by the pump (Note: 8 h = 480 min)?

Solution 5.2

$$V\,(m^3) = Q\,(\mathrm{l/min}) \times T\,(\mathrm{min}) \times \frac{1\ m^3}{10^3 l}$$

$$= 1.7\ \mathrm{l/min} \times 480\,\mathrm{min} \times \frac{1\ m^3}{10^3 l}$$

$$= 0.816\ m^3$$

Example 5.3

A 0.816 m^3 volume of air was sampled for respirable dust using a high-flow air sampling pump connected to a two-stage plastic cassette with a pre-weighed 37-mm, 5.0-µm pore size PVC filter and support pad medium in a Dorr-Oliver Cyclone. Laboratory analysis of the filter using gravimetry indicated that the filter pre-weight was 13.494 mg and post-weight was 16.524 mg. What was the concentration (mg/m^3) of respirable dust in the air during the sampling period?

Solution 5.3

$$\text{Respirable Dust (mg}/m^3) = \frac{\text{Post-weight (mg)} - \text{Pre-weight (mg)}}{\text{Air Volume (m}^3)}$$

$$= \frac{16.524 \text{ mg} - 13.494 \text{ mg}}{0.816 \text{ m}^3}$$

$$= \frac{3.03 \text{ mg}}{0.816 \text{ m}^3}$$

$$= 3.71 \text{ mg}/\text{m}^3$$

UNIT 5 CASE STUDY: SAMPLING AND ANALYSIS OF AIRBORNE RESPIRABLE PARTICULATES

OBJECTIVES

Upon completion of Unit 5, including sufficient reading and studying of this and related reference material, and review of the short case study and hypothetical industrial hygiene sampling and analytical data presented below, learners will be able to:

- Calculate the concentrations and time-weighted averages (TWAs).
- Interpret the data and results.
- Concisely summarize the information in a concise report.
- Retain the general scientific and technical principles and concepts related to the topic.

INSTRUCTIONS

After reading the short case study, locate and read the applicable NIOSH method for sampling and analysis (http://www.cdc.gov/niosh/nmam/nmammenu.html). Also, for more information about crystalline silica (SiO_2), including calculations of concentration, please refer to the OSHA site "Silica Advisor" (http://www.osha.gov/SLTC/silica_advisor/mainpage.html). Next, review the data shown in the tables below and perform the necessary calculations to complete the tables (assume normal temperature pressure). Review and interpret the results and compare calculated TWAs to applicable occupational exposure limits (e.g., Occupational Safety and Health Administration permissible exposure limits [OSHA-PELs], American Conference of Governmental Industrial Hygienists threshold limit values [ACGIH-TLVs], and Mine Safety and Health Administration threshold limit values [MSHA-TLVs]). Also, prepare responses to the questions that follow the data section. Finally, prepare a fictional, but representative, report following the outline or similar format shown in Appendix B.

CASE STUDY

Two masons were working together manually emptying 50-lb. bags of dry sand and mortar into a small mixer on-site in preparation for constructing a brick wall for a commercial establishment. The material was very powdery and generated visible airborne dust when poured and mixed. Although there was a slight wind, there remained a concern for workers' inhaling airborne fugitive emissions of dust. Accordingly, industrial hygiene integrated personal air sampling for respirable dust was conducted by a certified industrial hygienist (CIH). Following sampling, she prepared samples and field blanks prior to sending them to an American Industrial Hygiene Association (AIHA) accredited laboratory for analysis. Sampling and analysis for respirable dust and respirable quartz followed NIOSH Methods #0600 and #7500, respectively. It was shown that each respirable dust sample consisted of 8% quartz. Calibration, field, and remaining laboratory data are summarized in the tables below.

DATA AND CALCULATIONS

Table 5.2 Calibration and Field Data

Name of Personnel or Area	Sample No.	Analyte	Avg. Pre-sample Q (l/min)	Avg. Post-sample Q (l/min)	Avg. Q (l/min)	Sample Start Time	Sample Stop Time	Sample Time (min)	Air Volume Sampled (l)
Tim D.	0321	Respirable dust and quartz	Pump No. 221: 1.70	Pump No. 221: 1.72		0700	1100		
	0323					1101	1500		
Mike K.	0329	Respirable dust and quartz	Pump No. 225: 1.69	Pump No. 225: 1.72		0701	1102		
	0333					1103	1501		
Blank	0300	Respirable dust and quartz	N/A	N/A					

N/A = Not applicable.

Table 5.3 Lab Data

Name of Personnel or Area	Sample No.	Analytes	Pre-sample Weight (mg)	Post-sample Weight (mg)	Weight Respirable Dust (mg)	Percent Crystalline Quartz (%)	Air Volume Sampled (m³)	Conc. Respirable Dust (mg/m³)	Conc. Respirable Quartz Dust (mg/m³)
Tim D.	0321	Respirable dust and quartz	12.056	13.964					
	0323		11.999	14.004					
Mike K.	0329	Respirable dust and quartz	12.115	13.075					
	0333		12.104	13.604					
Blank	0300	Respirable dust and quartz	12.223	12.223	<LOD	N/A			

LOD = Limit of detection.

N/A = Not applicable.

Table 5.4 Calculated Time-Weighted Averages

Name of Personnel or Area	Sample No.	Analytes	8-h TWA Respirable Dust (mg/m³)	8-h TWA Respirable Quartz Dust (mg/m³)
Tim D.	0321 0323	Respirable dust and quartz		
Mike K.	0329 0333	Respirable dust and quartz		
Blank	0300	Respirable dust and quartz		

QUESTIONS

- Were the average pre- and post-sampling flow rates within ±5% of each other?
- Was the total sample volume collected within the min-max range specified in the applicable NIOSH method?
- What is the minimum number of field blanks that should be submitted to the laboratory, based on the number of samples collected and the applicable NIOSH method?
- What considerations or assumptions did you make when calculating the TWAs? For example, did you assume the workers took breaks (e.g., lunch) in areas outside the immediate area? Would you continue monitoring while workers are on breaks, even if outside the potential exposure area? If not, what if the actual sample collection period was less than 8 h (480 min)?
- What is the estimated limit of detection (LOD) for the analyte as per the applicable NIOSH method?
- How did the calculated TWAs compare to applicable occupational exposure limits, such as the OSHA-PEL, ACGIH-TLV, or MSHA-TLV?
- What was the basis for the established occupational exposure limits? (Learners and practitioners alike are strongly encouraged to obtain and read applicable information found in resources such as the NIOSH *Criteria Documents* and ACGIH *Documentation of the TLVs and BEIs*.)
- Are there other NIOSH or OSHA air sampling and analytical methods for the same agent? If so, what are the reference numbers and what are the differences relative to sample collection (e.g., medium used) and analysis (e.g., analytical instrument used)?

REPORT

Prepare a concise report (see Appendix B for outline of sample format) that includes typed text and tables (and figures if you desire). Your report needs to include a section on calibration (how you calibrate pump with representative media), sample collection (instruments, media, and NIOSH method used), and analysis (instrument and NIOSH method used). In other words, write your short field report so the reader knows what was done to prepare for sampling and analysis for respirable particulate (including crystalline silica). Also, summarize, interpret, and discuss the results; state your conclusions; and make applicable recommendations. Remember to add a bibliography at the end of the report citing any applicable book, journal, and Internet references that you consult. Finally, append a copy of the applicable NIOSH methods.

UNIT 6

Evaluation of Airborne Fibers as Asbestos: Integrated Personal and Area Monitoring Using an Air Sampling Pump with a Mixed Cellulose Ester Filter Medium

LEARNING OBJECTIVES

At the completion of Unit 6, including sufficient reading and studying of this and related reference material, learners will be able to correctly:

- Name, identify, and assemble the components of a common sampling train, including the specific sampling medium, used for conducting integrated sampling of airborne asbestos fibers.
- Name, identify, and assemble the applicable calibration train that uses a primary standard for measuring flow rate of a multi- or high-flow air sampling pump.
- Calibrate a multi- or high-flow air sampling pump, with the applicable representative sampling medium in-line, using a manual or electronic frictionless bubble-tube.
- Summarize the principles of sample collection of airborne asbestos fibers.
- Conduct integrated sampling of airborne asbestos fibers.
- Prepare samples for analysis of collected fibers using phase contrast microscopy.
- Name and identify the components of a phase contrast microscope.
- Name, identify, and assemble the components necessary for analysis of samples for asbestos fibers using a phase contrast microscope.
- Calibrate a Walton-Beckett graticule, positioned in the eyepiece of a phase contrast microscope, using a stage micrometer.
- Summarize the principles of sample analysis of filters used for the collection of airborne asbestos fibers.
- Conduct analysis of a sample for fibers using a phase contrast microscope.
- Perform applicable calculations and conversions related to pre- and post-sampling flow rate of an air sampling pump, average flow rate of an air sampling pump, sampling time, sampled air volume, number of fibers counted, and concentration of fibers in units of fibers per cubic meter (f/cm^3).
- Record all applicable calibration, sampling, and analytical data using calibration, field sampling, and laboratory analysis data forms.

OVERVIEW

There are situations when only knowing the concentration of inhalable, total, or respirable mass fractions of particulate in the air is insufficient since other airborne agents of concern and relevance may be present. If there is a real or potential source of fibrous dusts, such as asbestos, it is important to know what the concentration of these specific particulates is due to their inherent toxicity.

The morphology of fibers, unlike other classes of particles, is more columnar than sphere- or disk-like. As a result, the sampling and analysis involves methods that will assure accurate collection and quantification of fibers such as asbestos. Since fibers are particulates, they are collected on filters (as discussed earlier) for total and respirable particulate. Sampling involves the use of a different filter and cassette than those used to collect the more generic total or respirable particulates. In addition, the method of analysis involves microscopy instead of gravimetric analysis. Asbestos fibers are an example of a specific class of particulate air contaminants that require an analytical method that will both identify the specific analyte (fibers vs. nonfibers) and measure (count) the number of fibers collected to determine the airborne concentration.

PRECAUTIONS

Care must be taken not to collect too much particulate on the sampling filters. Since the analytical method requires microscopic examination of the filter for fibers and counting of those that meet specific criteria, it is necessary to have minimal interference from other particulates. An overloaded filter makes it difficult to count, since the fibers may be deposited in layers that appear as different viewing planes under the microscope. Accordingly, sampling flow rates and times should be determined, in part, based on estimations of concentrations of airborne particulate.

SAMPLING

A sample is collected using a 25-mm diameter mixed cellulose ester (MCE) filter with a 0.8-μm pore size. MCE filters are used because of the method of analysis for asbestos fibers. MCE filters can be made transparent in preparation for analysis so that individual fibers may be counted via microscopy. The filter is positioned on a cellulose support pad in a three-stage cassette. The cassette, however, differs from standard designs in that the center stage is a 50-mm extended cowl and only two stages (outlet and center) are used during sampling (Figure 6.1). The cowl permits more uniform collection and distribution of fibers across the face of the filter. The standard inlet port (third stage) of the cassette is not used during sampling because it can cause fibers to concentrate near the center of the filter and make it impractical to accurately count individual fibers. Care must be taken not to overload the filter for related reasons. The conductive extension cowl overcomes the effects of static electricity. The cowl creates a streamline effect, speeding the passage of the airstream to the filter face. This reduces the chance of fibers being electrostatically drawn to the sides of the cassette.

The cassette is assembled and capped on the inlet and outlet ports. When used for sampling, the plugs are removed from the cassette as is the third-stage cap. The open-faced cassette is attached to an air sampling pump using a length of flexible hose (Figure 6.2). A typical sampling rate is 2.0 l/min, but it ranges from 0.5 to 16 l/min depending on the purpose for sampling. For example, personal air samples are often collected at flow rates of 1.5 to 2.5 l/min, but area air samples are collected at substantially higher flow rates during asbestos-containing material abatement or post-abatement (i.e., clearance) activities. Following sampling, the cassette is

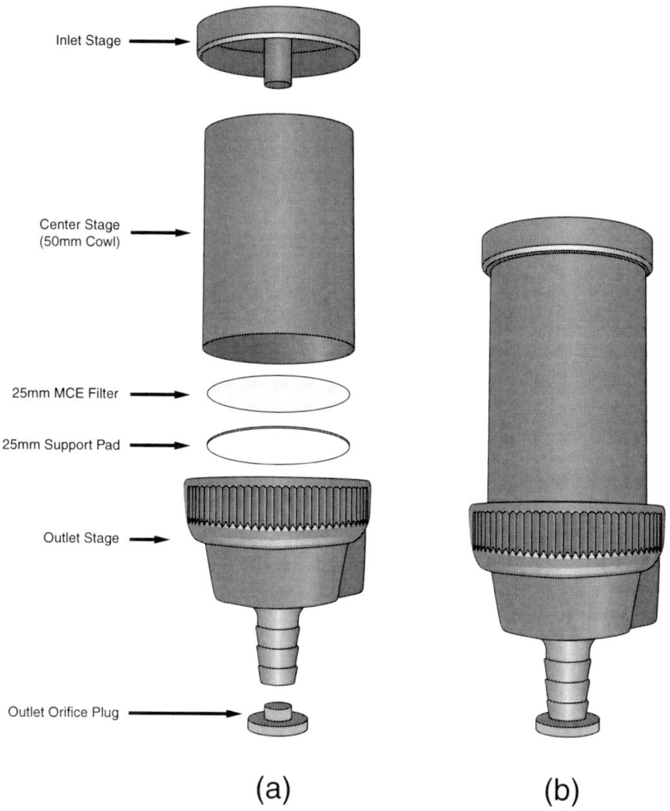

Figure 6.1 A 25-mm three-stage cassette with a 25-mm MCE filter and support pad. (a) Exploded view and (b) assembled view.

removed, the third-stage cap is replaced, and the open ends are closed with the plugs in the field. The sample is then transported to a laboratory for analysis. A brief outlined summary of sample collection follows.

(i) Calibration

- An air sampling pump is calibrated using a manual or electronic calibrator with representative sampling medium in-line (i.e., an assembled two-stage, open-faced cassette with a 50-mm extension cowl containing 25-mm, 0.8-μm pore size MCE filter and support) in-line and positioned in a 2-l calibration jar (Figure 6.3).
- Pre- (Q_{pre}) and post-sampling (Q_{post}) flow rates are determined by measuring the average time (T_{avg} [sec]) based on three trials ($\Sigma T_{1,2,3}/3$) for a bubble to traverse a specific volume (V [cm^3]) of the bubble-tube.

$$Q\ (\mathrm{l/min}) = \frac{\mathrm{V}\ (\mathrm{cm}^3)}{T_{avg}(\mathrm{sec})} \times \frac{60\ \mathrm{sec}}{1\ \mathrm{min}} \times \frac{1\ l}{10^3 cm^3} \tag{6.1}$$

- The average flow rate (Q_{avg}) is based on the average of pre- (Q_{pre}) and post-sampling (Q_{post}) flow rates.

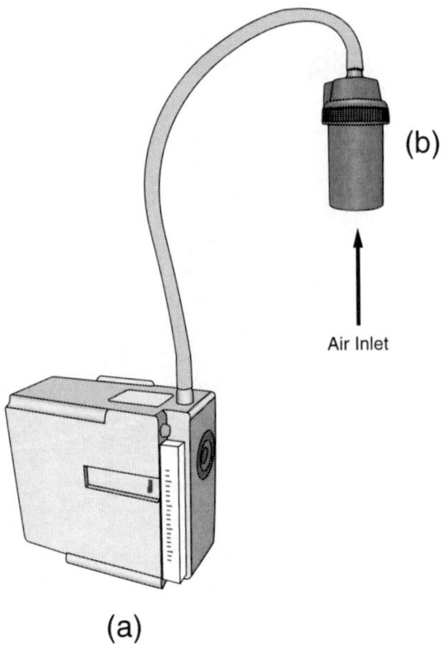

Figure 6.2 Sampling train for asbestos fibers composed of (a) a high-flow pump connected with flexible hose to (b) a 25-mm MCE filter in an open-faced, two-stage cassette.

$$Q_{avg} = \frac{Q_{pre} + Q_{post}}{2} \tag{6.2}$$

- The typical flow rate is 2 l/min and ranges from 0.5 to 16 l/min.

(ii) Preparation for Sampling

- An unweighed MCE filter plus support pad are inserted in a plastic cassette which is then assembled and labeled.
- The sampling train is assembled and consists of multi- or high-flow pump; 1/4-in. internal diameter (i.d.) flexible hose with clips; and a 25-mm two-stage plastic cassette with a 50-mm extension cowl containing a 25-mm, 0.8-µm pore size MCE filter on a cellulose support pad.

(iii) Conducting Sampling

- A labeled, open-faced (i.e., without third stage) plastic cassette is positioned with the opening facing downward and attached within the breathing zone, and the air sampling pump is clipped to the belt of the worker for personal sampling. The sampling train is positioned and secured in a specific location for area sampling.
- The pump is turned "ON" and start time is recorded.
- After a specified time, the pump is turned "OFF" and stop time is recorded.
- The third stage is placed on the plastic cassette and orifice plugs are inserted into the inlet and outlet ports. The sample is then transported to laboratory for analysis.

ANALYSIS

Analysis of the air sample typically involves either phase contrast microscopy (PCM) or transmission electron microscopy (TEM). TEM permits detailed examination and identification of

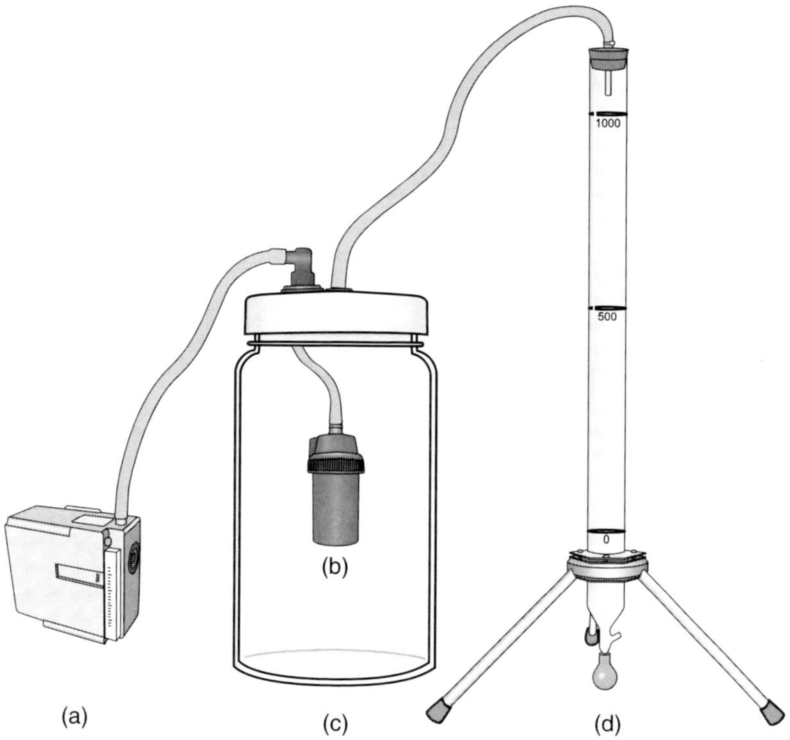

Figure 6.3 Calibration train for asbestos fibers composed of (a) a high-flow pump connected with flexible hose to (b) a 25-mm MCE filter in an open-faced two-stage cassette within a (c) calibration jar in-line with (d) a frictionless bubble-tube.

asbestos fibers. The method detects and measures both small and large fibers and permits differentiation between asbestos and non-asbestos fibers and types of asbestos (e.g., chrysotile vs. crocidolite). TEM is the most sensitive and accurate instrument for analyzing samples for asbestos fibers.

PCM is presently the most common analytical method for only counting fibers (Figures 6.4 and 6.5). Unlike TEM, PCM analysis is not used to identify the type of fiber. It is commonly used for samples collected in areas known to have sources of asbestos or friable asbestos-containing materials. The fibers counted on samples collected in source areas, based on relatively specific counting criteria, are reported as asbestos. To confirm the fiber identification, TEM (or less preferred scanning electron microscopy [SEM]) is required.

Samples are prepared for PCM analysis by cutting a segment of the MCE filter and mounting it on a glass microscopy slide (Figure 6.6). The originally white filter section is digested and collapsed on a microscope slide via acetone vapors yielding a transparent section. Subsequently, the transparent mounted filter section is examined using phase contrast illumination.

A Walton-Beckett graticule is positioned in the ocular of the microscope (Figure 6.7). While looking through the ocular of the microscope, the graticule appears as a grid superimposed over the microscope viewing field of the sample. The graticule grid, which is calibrated using a stage micrometer, is used for measuring the length, width, and number of fibers. The standard diameter is 100 μm (0.1 mm^2), and the cross-sectional area (based on πr^2) is approximately 0.00785 mm^2. The interior of the circle is the counting field of the Walton-Beckett graticule. The crosshairs and the rectangles around the perimeter permit measurement of fibers. Any particle having a length to width ratio greater than 3:1 and a length of 5 μm or greater is counted as a fiber. Samples are viewed at 400× (i.e., 10× ocular lens × 40× objective lens) magnification power, and results are presented as the number of fibers per cubic centimeter of air (f/cm^3).

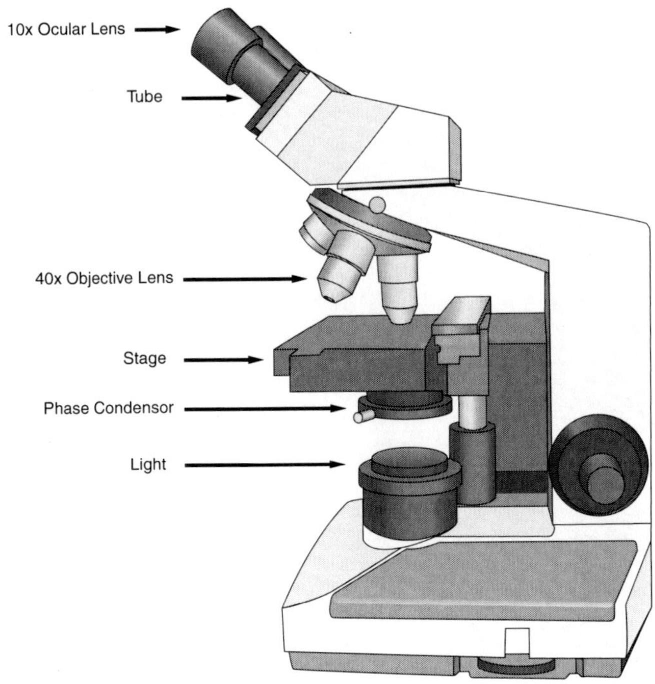

Figure 6.4 Components of a phase contrast microscope.

Again it should be noted that PCM is a counting method and does not measure the specific properties of a substance. It is not inherently specific for asbestos. All particles meeting the criteria described (length-to-width ratio and length) are counted as fibers that are possibly asbestos. Smaller fibers that may be present are not counted.

A detailed description of these instruments is beyond the scope of this book. A brief outlined summary of the components of a phase contrast microscope and the principles using phase contrast illumination follows.

(i) Components

- Ocular/eyepiece: contains a viewing lens with 10× magnification power; a Walton-Beckett graticule is positioned within the ocular below the lens
- Body tube: connects the ocular lens to the objective lens and serves as a conduit for transmitted light
- Objective: lens with 40× magnification power positioned directly over the sample; total magnification equals the product of magnification powers for ocular and objective lenses (e.g., 10× ocular lens multiplied by 40× objective lens yields 400× magnification power)
- Stage: platform where sample mounted on microscope slide is positioned under the objective lens
- Phase condenser: positioned below the stage and above a light source; focuses and adjusts the phase of the wavelength of light projected to sample
- Light: positioned below the condenser and at the base of the microscope; provides necessary illumination of the sample

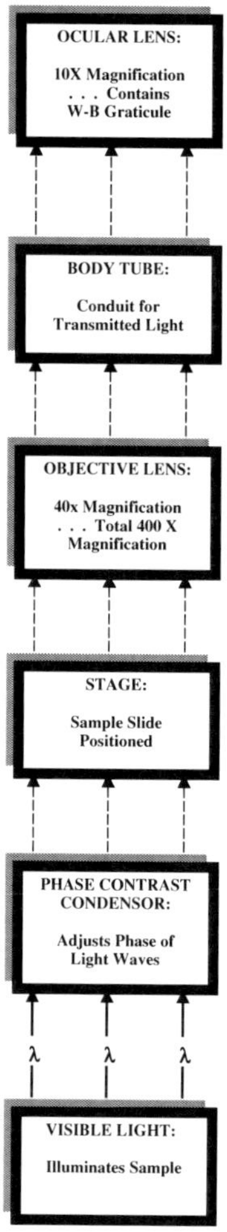

Figure 6.5 Schematic of a phase contrast microscope.

(ii) Principle of Operation

- A wavelength of visible light is projected from the light source to the sample and through objective and ocular lenses. Contrast is created due to the difference between intensity of light traveling through air and diffracted by the fibers deposited on sample.
- Light waves passing undisturbed around the sample are undiffracted. Light waves disturbed as a result of the sample are diffracted and, accordingly, out of phase relative to the undiffracted light. The diffraction is due to the thickness and refractive index of the sample.
- Interaction between diffracted with undiffracted light results in destructive interference and related differences in intensity. The intensity differences create light and dark contrast.
- The contrast or magnified image would be relatively poor when analyzing samples for mineral fibers such as asbestos. The phase contrast condenser, however, alters and focuses the phase and results in enhanced contrast for viewing fibers.

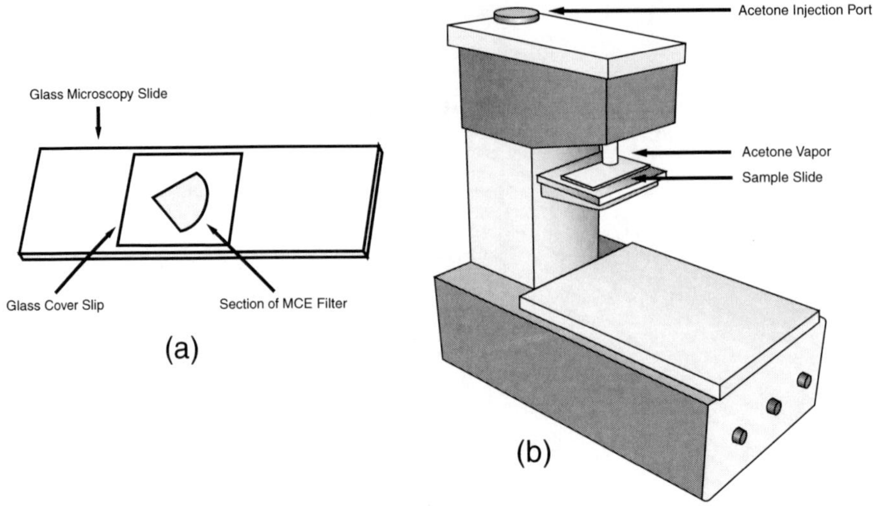

Figure 6.6 (a) Section of MCE sample filter mounted on a standard microscope slide. (b) Sample slide positioned on acetone hot-block.

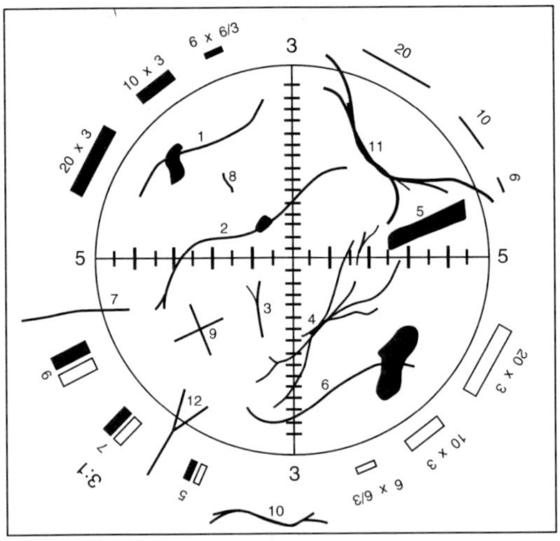

Figure 6.7 Viewing area of a Walton-Beckett graticule.

(iii) Determination of Concentration of Fibers

- Calibrate a Walton-Beckett graticule to determine its field area (A). The cross-sectional or field area should be approximately 0.00785 mm^2.
- Calculate the number of fibers per square millimeter (mm^2) of counting area (fiber density [E]) by counting fibers on sample filter (F) per total fields counted (n_f), subtracting fibers counted on blank filter (B) per total fields (n_b), and dividing by the counting area of the Walton-Beckett graticule (A_f: 0.00785 mm^2).

$$E\,(\mathrm{f/mm^2}) = \frac{\left(\frac{F}{n_f} - \frac{B}{n_b}\right)}{A_f} \qquad (6.3)$$

- Determine the sampled volume of air by multiplying sampling time (T [min]) by average sampling flow rate (Q [l/min]) and express volume in liters.

$$\text{Air Volume } (l) = T\,(\text{min}) \times Q_{avg}\,(\text{l / min}) \qquad (6.4)$$

- Determine the concentration of fibers by multiplying the fiber density (E [f/mm^2]) by the effective collection area of the 25-mm filter (A_c: 385 mm^2) and dividing by the volume sampled (V [l]).

$$Fibers\,(\text{f / cm}^3) = \frac{E\,(\text{f / mm}^2) \times A_c(mm^2)}{\text{V (l)} \times \frac{10^3\ cm^3}{l}} \qquad (6.5)$$

UNIT 6 EXERCISE

OVERVIEW

The exercise will provide the fundamental concepts for conducting integrated sampling for fibers using a filter medium. In addition, a common and related analytical method using phase contrast microscopy is introduced. The exercise will focus on sampling and analysis of asbestos fibers. The material and methods are based on a modification of National Institute for Occupational Safety and Health (NIOSH) Method #7400. Place a check mark (✔) in the open box (☐) when you have obtained applicable material and completed the steps for sampling and analytical methods.

MATERIAL

1. Calibration of Air Sampling Pump

- ☐ Multi- or high-flow air sampling pump
- ☐ 25-mm, 0.8-µm pore size MCE filter on a cellulose support pad in an assembled 25-mm two-stage, open-faced plastic cassette with a 50-mm extension cowl
- ☐ Two 2- to 3-ft lengths of ¼-in. i.d. flexible hose
- ☐ 2-l calibration jar
- ☐ Manual or electronic frictionless bubble-tube calibrator
- ☐ Soap solution
- ☐ Stopwatch (if manual calibrator is used)
- ☐ Calibration data form (Figure 6.8)

2. Sampling for Asbestos Fibers

- ☐ 25-mm, 0.8-µm pore size MCE filter with a cellulose support pad in assembled three-stage plastic cassette
- ☐ Orifice plugs
- ☐ Multi- or high-flow air sampling pump
- ☐ One 2- to 3-ft length of ¼-in. i.d. flexible hose
- ☐ Cellulose cassette bands or tape to secure the stages of the cassette (optional)
- ☐ Field sampling data form (Figure 6.9)

Calibration Data Form:
Low-Flow, High-Flow, and Multi-Flow Air Sampling Pumps

Name and Location of Calibration: ____________________

Calibration Conducted By: ____________________ Date Calibration Conducted: ________

Calibration Instruments (Type/Manufacturer/Model): ____________________

Air Sampling Pump
(Type/Manufacturer/Model): ____________________

Collection Medium In-Line: ____________________ Calibrator Volume: ________cc

Air Temperature: ____°C Air Pressure: ____mm Hg Relative Humidity: ____%

Pump No.	Pre-Sampling Calibration Flow Rate (Q_{pre})					Post-Sampling Calibration Flow Rate (Q_{post})					Average Flow Rate
	Time (sec)				Q_{pre}	Time (sec)				Q_{post}	Q_{avg}
	T_1	T_2	T_3	T_{avg}	L/min	T_1	T_2	T_3	T_{avg}	L/min	L/min

Calibration Notes:

Figure 6.8 Calibration data form for high-flow air sampling pump.

Field Monitoring Data Form:
Integrated Monitoring for Asbestos (as Total Fibers)

Facility Name and Location: ______________________ **Contaminant Sampled:** ☐ **Total Fibers**

☐ **Asbestos (form?)** ________

Monitoring Conducted By: ______________________

Collection Medium: ______________________

Date Monitoring Conducted: ______________________

Method(s)(Source/Number/Name): ______________________

Monitoring Instruments (Type/Manufacturer/Model): ______________________

Air Temperature: ____°C **Air Pressure:** ____mm Hg **Relative Humidity:** ____%

Identification of Personnel or Area	Field Sample	Pump No.	Pre-Sample Calibration Flow Rate	Post-Sample Calibration Flow Rate	Average Flow Rate (Q)	Sample Start-Time	Sample Stop-Time	Total Sample Time (T)	Volume Sampled (Vol)

Field Notes:

Figure 6.9 Field sampling data form for asbestos fibers.

3. Analysis of Sample for Asbestos Fibers

- ☐ Phase contrast microscope
- ☐ Walton-Beckett graticule
- ☐ Stage micrometer
- ☐ Microscope slides
- ☐ Scalpel, probe, and forceps
- ☐ Waterproof marker
- ☐ Cover slips
- ☐ Acetone
- ☐ Acetone hot block
- ☐ Triacetin
- ☐ 1-cm^3 syringes
- ☐ Laboratory analysis data form (Figure 6.10)

METHOD

1. Pre-sampling Calibration of Air Sampling Pump

- ☐ Remove pump from the charger, turn "ON," and allow it to run for 10 min.
- ☐ Complete a calibration data form, remembering to record name, location, data, air temperature, air pressure, relative humidity, pump manufacturer and model, calibration apparatus, times for three trials, and average flow rate for pre- and post-sampling calibrations.
- ☐ Assemble a representative sampling medium for use during calibration of the air sampling pump. This should consist of a 25-mm three-stage plastic cassette with a 50-mm extension cowl, containing a 25-mm, 0.8-μm pore size MCE filter and cellulose support pad.
- ☐ Connect the flexible hose from inside the calibration jar to the orifice of the outlet stage of the cassette.
- ☐ Remove the inlet (third) stage of the cassette.
- ☐ Place the two-stage, open-faced cassette into the calibration jar and tighten the lid securely. Wrap plastic tape around the perimeter of the sealed lid to reduce leakage of outside air.
- ☐ Connect a 2- to 3-ft length of flexible hose to the air sampling pump.
- ☐ Connect the other end of the flexible hose to tube projecting out of the lid of the calibration jar so that there is a connection between the two-stage, open-faced cassette and lid to pump via flexible hose.
- ☐ Connect another 2- to 3-ft length of flexible hose to the other tube that projects out of the calibration jar.
- ☐ Connect the remaining open end of the flexible hose to a manual or automatic frictionless bubble-tube calibration device. This constitutes the calibration train.
- ☐ Aspirate the bubble solution into the calibration device to form a soap film (bubble).
- ☐ Record the time it takes the soap film (bubble) to traverse a volume of 500 to 1000 ml.
- ☐ If a manual bubble tube is used, calculate the flow rate as shown below. Otherwise read the flow rate directly from the electronic bubble-tube calibrator.
- ☐ If the flow rate is not 2.0 l/min ±5%, adjust the flow control on the pump and repeat calibration procedure until desired flow rate is achieved.
- ☐ Once the desired flow rate is achieved (in this example 2.0 l/min ±5%), repeat the calibration check two to three times and calculate the "average pre-sampling flow rate."
- ☐ Disconnect the pump from the remainder of the calibration apparatus and turn "OFF." The calibrated air sampling pump is ready for use.

Laboratory Analysis Data Form:
Phase Contrast Microscopy Analysis of Fibers

Monitored Facility Name and Location: ______________ Contaminant Sampled: ______________

Monitoring Conducted By: ______________ Date Monitoring Conducted: ______________

Method (Source/Number/Name): ______________

Laboratory Facility Name and Location: ______________ Contaminant Analyzed: ______________

Analysis Conducted By: ______________ Date Analysis Conducted: ______________

Filter Collection Medium (Type/Manufacturer): ______________ W-B Graticule Calibration: ____ mm ____ mm^2

Phase Contrast Microscope (Type/Manufacturer/Model): ______________

PCM Calibration: ______________

Air Temperature: ____°C Air Pressure: ____mm Hg Relative Humidity: ____%

Laboratory Sample Identification No.	Field Sample Identification No.	Sample: Fibers Counted (F) per Fields (n_f)	Blank: Fibers Counted (B) per Fields (n_b)	Graticule Counting Area (A_f) (mm^2)	Fiber Density (E) (f/mm^2)	Effective Collection Area (A_c) Filter (mm^2)	Average Flow Rate (Q) (L/min)	Total Sample Time (T) (min)	Volume Sampled (Vol) (L)	Conc. Fibers (f/cc)
		F =	B =							
		n_f =	n_b =							

Fiber Count (100 Fields)

Laboratory Notes:

Figure 6.10 Laboratory analysis data form for phase contrast microscopy analysis of asbestos fibers.

2. Preparation of Sample Media

- ☐ Using a forceps, place an unweighed cellulose support pad into the outlet stage (first stage) of the three-stage cassette.
- ☐ Using a forceps, gently remove an MCE filter from its package.
- ☐ Place the unweighed MCE filter on the support pad.
- ☐ Position (sometimes screw) the 50-mm extension cowl into the outlet stage to secure the filter and support pad.
- ☐ Place the inlet stage onto the cowl.
- ☐ Press the three stages of the cassette together by using your palm and a flat rigid plastic or metal plate to apply pressure.
- ☐ Place the orifice plugs in the inlet and outlet stages of the cassette.
- ☐ Place a wet cellulose band around the periphery of the first and second stages of the cassette and allow to dry and shrink. Tape may be used instead (optional).
- ☐ Label the cassette with a sample identification number.

3. Air Sampling for Fibers as Asbestos

- ☐ Get approval to conduct sampling in an area where airborne levels of asbestos fibers are likely to be detectable and measurable and perform sampling accordingly. Alternatively, conduct a simulated sampling exercise and obtain a sample spiked with asbestos fibers from your instructor for subsequent analysis.
- ☐ Complete a field sampling data form, remembering to record name, location, person or area sampled, date, temperature, pressure, analyte (e.g., "asbestos fibers"), air sampling pump manufacturer and model, pump identification, sampling medium, sample identification, flow rate, start and stop times for sampling, sampling duration, and air volume sampled.
- ☐ Turn "ON" pump and let run for 5 min.
- ☐ Place and secure the pump in an area where sample will be collected.
- ☐ Connect a flexible hose to pump.
- ☐ Remove the orifice plug from the outlet stage and the entire inlet stage of the cassette.
- ☐ Attach the two-stage, open-faced cassette to flexible tubing that is connected to pump.
- ☐ Record the sample number and start time and allow sample to run for at least 2 h depending on conditions.
- ☐ After an acceptable sampling period has elapsed, turn "OFF" pump and record stop time.
- ☐ Remove the cassette from the sampling train, insert the orifice plug into the outlet stage and replace the inlet stage (including orifice plug). Transport sample (i.e., now closed-face, three-stage cassette) to laboratory.

4. Post-sampling Calibration of Air Sampling Pump

- ☐ Repeat steps summarized in Method #1.
- ☐ Determine average sampling flow rate by averaging the pre- and post-sampling flow rate and express average flow rate in liters per minute.

5. Analysis of Sample for Fibers as Asbestos

- ☐ Remove the cellulose band from around the periphery of the cassette.
- ☐ Disassemble the cassette by carefully removing the second and third stages to expose the filter.
- ☐ Using a probe and scalpel, slice approximately a 20 to 25% pie-shaped piece of the filter.
- ☐ Using a probe, gently push through the outlet stage orifice so that the support pad and filter are elevated and accessible for grasping with a forceps.
- ☐ Grasp the section of the filter and place on a clean, labeled microscopy slide.

- ☐ Using a waterproof marker, outline the perimeter of the filter section from underneath the microscope slide.
- ☐ Place the slide with filter section on the hot block.
- ☐ Gently inject 0.2 ml acetone into the hot block until filter section changes from white to colorless or transparent. The filter wedge is now considered cleared.
- ☐ Add one drop (3 to 3.5 ml) of triacetin to the cleared filter wedge, using the outlined tracing you drew underneath the slide as a guide for location, and place cover slip on top.
- ☐ Wait at least 15 min prior to analysis.
- ☐ Turn "ON" microscope and assure that the Walton-Beckett graticule is in one of the oculars.
- ☐ Adjust the microscope to 400× power and assure that the optics are set and focused according to specifications established by the manufacturer and the method.
- ☐ Place the stage micrometer on the stage of the microscope and focus microscope.
- ☐ While continuing to look through the ocular, align the peripheral edge of the circle of the graticule with a demarcation of the micrometer (each mark is 10 mm).
- ☐ Measure field diameter (D) of the Walton-Beckett graticule with the stage micrometer. The diameter should equal 100 ±2 μm; divide by 10^3 to convert micrometers to millimeters (i.e., 100 μm = 0.1 mm).
- ☐ Calculate field area (A) which should equal approximately 0.00785 mm^2.

$$
\begin{aligned}
A &= \pi\left(\frac{D}{2}\right)^2 \\
&= \pi\left(\frac{0.1\ mm}{2}\right)^2 \\
&= 0.00785\ \mathrm{mm}^2
\end{aligned}
\tag{6.6}
$$

- ☐ Record graticule calibration data on a laboratory data analysis form.
- ☐ Remove the stage micrometer and place the microscope slide sample centered on the stage of the calibrated microscope under the objective lens. Focus the microscope on the plane of the filter.
- ☐ Start counting from the tip of the filter and progress along a radial line to the outer edge. Shift up or down on the filter, and continue in the reverse direction. Select graticule fields randomly by looking away from the eyepiece briefly while advancing the manual stage. Ensure that, as a minimum, each analysis covers one radial line from the filter center to the outer edge of the filter.
- ☐ Count only fibers >5-μm long with a length-to-width ratio ≥3:1.
- ☐ Count each fiber that falls entirely within the graticule area as one, provided that the fiber meets criteria established above (i.e., fiber is longer than 5 μm and ≥3:1 length-to-width ratio). Count fibers (that meet the same criteria) as 0.5 if only one end is positioned within the graticule area. Count bundles of fibers as one. Do not count any other fibers, including those that cross the graticule area more than once.
- ☐ When an agglomerate covers one-sixth or more of the graticule field, reject the graticule field and select another. Do not report rejected fields in the total number counted.
- ☐ Count enough graticule fields to yield 100 fibers. Count a minimum of 20 graticule fields and stop at 100 graticule fields, regardless of count.
- ☐ Calculate concentration of asbestos fibers in units of fibers per cubic meter.
- ☐ Complete a laboratory analysis data form, remembering to record name, location, date, air temperature, air pressure, analytes (fibers, [e.g., asbestos]), phase contrast microscope manufacturer and model, sample identification, air volume sampled, and concentration.

UNIT 6 EXAMPLES

Example 6.1

A high-flow air sampling pump was calibrated using a manual frictionless bubble-tube. What was the flow rate (Q) in liters per minute (l/min) if it took 12 sec (T) for the bubble to traverse a volume (V) of 500 cm^3?

Solution 6.1

$$Q\ (\mathrm{l/min}) = \frac{\mathrm{V}\ (cm^3)}{T\ (\mathrm{sec})} \times \frac{60\ \mathrm{sec}}{\mathrm{min}} \times \frac{1\ \mathrm{l}}{10^3\ cm^3}$$

$$= \frac{500\ cm^3}{12\ \mathrm{sec}} \times \frac{60\ \mathrm{sec}}{\mathrm{min}} \times \frac{1\ \mathrm{l}}{10^3\ cm^3}$$

$$= 2.5\ \mathrm{l/min}$$

Example 6.2

A sample was collected for 3 h (T [min]) using a high-flow sampling pump at a flow rate (Q) of 2.5 l/min. What was the volume (V [m^3]) of air sampled by the pump (Note: 3 h = 180 min)?

Solution 6.2

$$\mathrm{V}\ (m^3) = Q\ (\mathrm{l/min}) \times T\ (\mathrm{min}) \times \frac{1\ m^3}{10^3 l}$$

$$= 2.5\ \mathrm{l/min} \times 180\ \mathrm{min} \times \frac{1\ m^3}{10^3 l}$$

$$= 0.450\ m^3 \text{ or } 450\ \mathrm{l}$$

Example 6.3

A 450-L volume of air was sampled for asbestos fibers using a high-flow air sampling pump connected to a two-stage, open-faced cassette with a 25-mm, 0.8-μm pore size MCE filter and support pad medium. Laboratory analysis of the filter using phase contrast microscopy indicated a count of 100 fibers/42 fields. What was the concentration (f/cm^3) of asbestos fibers in the air during the sampling period? (Assume graticule area = 0.00753 mm^2, filter area = 385 mm^2, and Blank = 0 fibers/100 fields.)

Solution 6.3

$$E\ (\text{f} / \text{mm}^2) = \frac{\frac{100\ fiber}{42\ field} - \frac{0\ fiber}{100\ field}}{0.00753\ \ mm^2}$$

$$= 316.2\ \ \text{f} / \text{mm}^2$$

$$Fibers\ \ (\text{f} / \text{cm}^3) = \frac{316.2\ \text{f} / \text{mm}^2 \times 385\ \ mm^2}{450\ \ l \times \frac{10^3\ cm^3}{1\ l}}$$

$$= 0.27\ \text{f} / \text{cm}^3$$

UNIT 6 CASE STUDY: SAMPLING AND ANALYSIS OF AIRBORNE ASBESTOS FIBERS

OBJECTIVES

Upon completion of Unit 6, including sufficient reading and studying of this and related reference material, and review of the short case study and hypothetical industrial hygiene sampling and analytical data presented below, learners will be able to:

- Calculate the concentrations and time-weighted averages (TWAs).
- Interpret the data and results.
- Concisely summarize the information in a concise report.
- Retain the general scientific and technical principles and concepts related to the topic.

INSTRUCTIONS

After reading the short case study, locate and read the applicable NIOSH method for sampling and analysis (http://www.cdc.gov/niosh/nmam/nmammenu.html). Next, review the data shown in the tables below and perform the necessary calculations to complete the tables (assume normal temperature pressure). Review and interpret the results and compare calculated TWAs to applicable occupational exposure limits (e.g., Occupational Safety and Health Administration permissible exposure limits [OSHA-PELs], American Conference of Governmental Industrial Hygienists threshold limit values [ACGIH-TLVs], and NIOSH recommended exposure limits [NIOSH-RELs]). Also, prepare responses to the questions that follow the data section. Finally, prepare a fictional, but representative, report following the outline or similar format shown in Appendix B.

CASE STUDY

Integrated personal air monitoring for asbestos fibers was conducted by a certified industrial hygiene technician (CIHT) in a truck maintenance shop. The exposure group of interest was the service technicians that removed and replaced brake pads on numerous types of construction vehicles (e.g., dump trucks). The brake pads were known to contain a substantial amount of asbestos. Insignificant local exhaust ventilation was provided. Air sampling and analysis for asbestos (as total fibers) followed NIOSH Method #7400. Calibration, field, and laboratory data are summarized in the tables below.

DATA AND CALCULATIONS

Table 6.1 Calibration and Field Data — Asbestos Fibers

Name of Personnel or Area	Sample No.	Analyte	Avg. Pre-sample Q (l/min)	Avg. Post-sample Q (l/min)	Avg. Q (l/min)	Sample Start Time	Sample Stop Time	Sample Time (min)	Air Volume Sampled (l)
Gary M.	0052	Fibers	Pump No. 133: 2.25	Pump No. 133: 2.25		1503	1930		
	0055					1931	2300		

Table 6.2 Lab Data — Asbestos Fibers

Name of Personnel or Area	Sample No.	Analyte	Fibers Counted Per Total Fields (f/field)	Fiber Density (f/mm²)	Air Volume Sampled (cm³)	Conc. Fibers (f/cm³)
Gary M.	0052	Fibers	100/67			
	0055		82/100			

Lab Notes: (1) Calibration of the Walton-Beckett graticule using a stage micrometer indicated that the measured diameter was 99.0 μm. (2) Assume total exposed surface of 25-mm diameter MCE filter during sampling was 385 mm². (3) Assume all field blanks were lower than the limit of detection (LOD).

Table 6.3 Calculated Time-Weighted Averages

Name of Personnel or Area	Sample No.	Analyte	8-h TWA Asbestos Fibers (f/cm³)
Gary M.	0052 0055	Fibers	
Blank	0500	Fibers	

QUESTIONS

- Were the average pre- and post-sampling flow rates within ±5% of each other?
- Was the total sample volume collected within the min-max range specified in the applicable NIOSH method?
- What is the minimum number of field blanks that should be submitted to the laboratory, based on the number of samples collected and the applicable NIOSH method?
- What considerations or assumptions did you make when calculating the TWAs? For example, did you assume the workers took breaks (e.g., lunch) in areas outside the immediate area? Would you continue monitoring while workers are on breaks, even if outside the potential exposure area? If not, what if the actual sample collection period was less than 8 h (480 min)?
- What is the estimated limit of detection (LOD) for the analyte as per the applicable NIOSH method?
- How did the calculated TWAs compare to applicable occupational exposure limits, such as the OSHA-PEL, ACGIH-TLV, and NIOSH-REL?

- What was the basis for the established occupational exposure limits? (Learners and practitioners alike are strongly encouraged to obtain and read applicable information found in resources such as the NIOSH *Criteria Documents* and ACGIH *Documentation of the TLVs and BEIs*.)
- Are there other NIOSH or OSHA air sampling and analytical methods for the same agent? If so, what are the reference numbers and what are the differences relative to sample collection (e.g., medium used) and analysis (e.g., analytical instrument used)?

REPORT

Prepare a concise report (see Appendix B for outline of sample format) that includes typed text and tables (and figures if you desire). Your report needs to include a section on calibration (how you calibrate pump with representative media), sample collection (instruments, media, and NIOSH method used), and analysis (instrument and NIOSH method used). In other words, write your short field report so the reader knows what was done to prepare for sampling and analysis for airborne asbestos fibers particulate (including crystalline silica). Also, summarize, interpret, and discuss the results; state your conclusions; and make applicable recommendations. Remember to add a bibliography at the end of the report citing any applicable book, journal, and Internet references that you consult. Finally, append a copy of the applicable NIOSH methods.

UNIT 7

Evaluation of Airborne Metal Dusts and Fumes: Integrated Personal and Area Monitoring Using an Air Sampling Pump with Mixed Cellulose Ester Filter Medium

LEARNING OBJECTIVES

At the completion of Unit 7, including sufficient reading and studying of this and related reference material, learners will be able to correctly:

- Name, identify, and assemble the components of a common sampling train, including the specific sampling medium, used for conducting integrated monitoring of airborne metal dust and fume.
- Name, identify, and assemble the applicable calibration train that uses a primary standard for measuring flow rate of a multi- or high-flow air sampling pump.
- Calibrate a multi- or high-flow air sampling pump, with the applicable representative sampling medium in-line, using a manual or electronic frictionless bubble-tube.
- Summarize the principles of sample collection of airborne metal dust and/or fume.
- Conduct integrated monitoring of airborne metal dust and fume.
- Prepare samples for analysis of collected particulate using atomic absorption or inductively coupled plasma emission spectroscopy.
- Name and identify the components of an atomic absorption spectrometer.
- Name, identify, and assemble the components necessary for analysis of samples for metal dust and fume using an atomic absorption spectrometer.
- Calibrate an atomic absorption spectrometer and establish a standard curve using known concentrations of a metal analyte.
- Summarize the principles of sample analysis of filters used for collection of airborne metal dust and fume.
- Conduct analysis of a sample for metal dust and/or fume using an atomic absorption spectrometer.
- Perform applicable calculations and conversions related to pre- and post-sampling flow rates of an air sampling pump, average flow rate of an air sampling pump, sampling time, sampled air volume, weight of metal dust and fume collected, and concentration of metal dust and/or fume in units of milligrams per cubic meter (mg/m^3).
- Record all applicable calibration, sampling, and analytical data using calibration, field monitoring, and laboratory analysis data forms.

OVERVIEW

Metallic dusts and fumes are common particulates in the occupational environment. Metals are examples of a class of toxic elements and compounds for which many specific occupational exposure limits for airborne levels have been established. If there is a real or potential source of metallic dusts or fumes, it is important to know what the concentration of specific metallic particulates is due to the toxicity associated with these compounds. For example, lead dust is considered highly toxic relative to nuisance dust or particulates not otherwise classified, and, accordingly, occupational exposure limits are presently 200 to 300 times lower.

Although there are similarities in the monitoring methods, overall collection and measurement of airborne metal dusts and fumes are more complicated when compared to sampling and analysis of the more generic total and respirable particulates. Nonetheless, metals are separated from a contaminated airstream using filtration that involves electrostatic attraction, interception, and impaction of metal particles with a filter medium.

FILTER MEDIUM FOR METAL DUSTS AND FUMES

The sample medium is a 37-mm diameter mixed cellulose ester (MCE) with a 0.8-µm pore size. MCE filters are used because, relative to polyvinyl chloride (PVC) filters, they hydrolyze (dissolve) in acid and ash easily — characteristics that are important during post-sampling analysis of samples for metals. As discussed previously regarding the PVC filters, the diameter (37 mm) of the MCE filter provides cross-sectional surface area for deposition and collection of particulate. The pore size (0.8 µm) of an MCE filter for sampling metal dusts and fumes, however, is much less than the pore size (5 µm) of the PVC filter used to collect total and respirable particulates. The smaller pore size filter is warranted since typically the variety of particle sizes is lower and the particle diameters smaller for metal dusts and especially fumes relative to total dusts. The mechanism of collection for PVC and MCE filters, however, is basically the same. The pore size restricts particulates with diameters greater than the pore diameter to collect only on the surface of the filter. Particulates with diameters less than the pore size, however, are also collected. These particles are collected within the pores due to an electrostatic attraction between the particles and the filter. In addition, the pores are tapered and allow interception and impaction of particulates within the medium. Collection efficiency also increases as particulates accumulate on the MCE filter.

The assembly of the MCE filter medium and cassette for metal dust and fume sampling is the same procedure as described for total particulate sampling, except for the use of different filters. The MCE filter medium is positioned on a cellulose support pad and secured within a 37-mm three-stage plastic cassette (Figure 7.1). The outlet stage holds the support pad and the filter and contains an orifice (hole) in the center which serves as the outlet port for sampled air to exit the cassette. The middle stage is a ring-shaped spacer or cowl which increases the depth or height of the cassette to facilitate more even deposition of the particulates from the air as it enters the cassette. The inlet stage contains a hole in the center which serves as the inlet port for sampled air to enter the cassette.

SAMPLING

The sampling method for metallic particulate is similar to those methods used to collect total particulate air contaminants. The most significant difference is that sampling for metals involves the use of a different filter medium. In addition, sample preparation does not involve gravimetry to pre-weigh the filter, since methods for analysis of the contaminants are different.

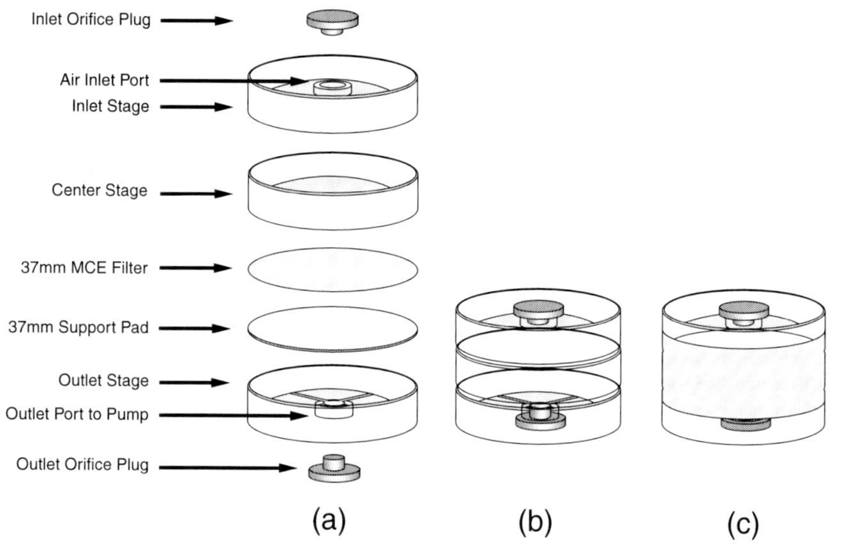

Figure 7.1 A 37-mm three-stage cassette with a 37-mm MCE filter and support pad. (a) Exploded view, (b) assembled view, and (c) assembled view secured with shrink band or tape.

Sample preparation simply involves positioning the MCE filter within and assembling the cassette. Following assembly, plastic plugs are inserted at the inlet and outlet ports or orifices. When used for sampling, the orifice plugs are removed and the cassette is attached to an air sampling pump using a length of 1/4 in. internal diameter (i.d.) flexible hose (Figure 7.2). Typical sampling rates range from 1 to 3 l/min. Following sampling, the cassette is removed and plugs are reinserted in the open orifices in the field. The sample is then transported to a laboratory for analysis.

A brief outlined summary of sample collection follows.

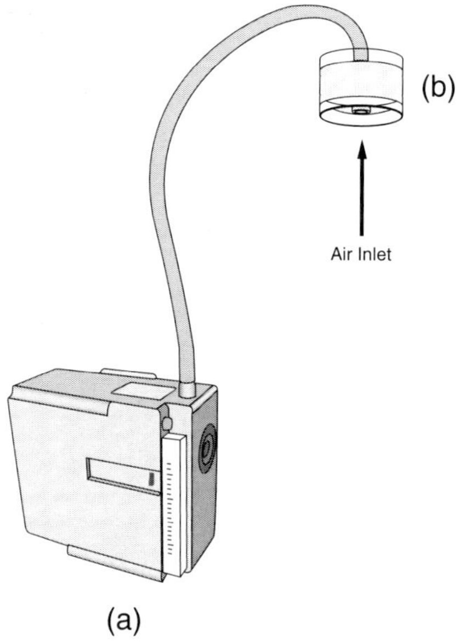

Figure 7.2 Sampling train for metal dust and fume composed of (a) a high-flow pump connected with flexible hose to (b) a 37-mm MCE filter in a three-stage cassette.

(i) Calibration

- An air sampling pump is calibrated pre- and post-sampling to adjust or determine flow rate using a manual or electronic calibrator with representative media (i.e., an assembled three-stage cassette containing a 37-mm, 0.8-µm pore size MCE filter and support pad) in-line (Figure 7.3).
- Pre- (Q_{pre}) and post-sampling (Q_{post}) flow rates are determined by measuring the average time (T_{avg} [sec]) based on three trials ($\Sigma T_{1,2,3}/3$) for a bubble to traverse a specific volume (V [cm^3]) of the bubble-tube.

$$Q\,(\mathrm{l/min}) = \frac{\mathrm{V\ (cm^3)}}{T_{avg}(\mathrm{sec})} \times \frac{60\ \mathrm{sec}}{1\,\mathrm{min}} \times \frac{1\,l}{10^3 cm^3} \tag{7.1}$$

- Average flow rate (Q_{avg}) is based on the average of pre- (Q_{pre}) and post-sampling (Q_{post}) flow rates.

$$Q_{avg} = \frac{Q_{pre} + Q_{post}}{2} \tag{7.2}$$

- Typical flow for sampling metal particulates are 1 to 3 l/min.

(ii) Preparation for Monitoring

- An unweighed MCE filter plus unweighed support pad are inserted in a plastic cassette which is then assembled and labeled.
- A sampling train is assembled and consists of a multi- or high-flow pump; a ¼-in. i.d. flexible hose with clips; and a 37-mm three-stage plastic cassette containing an unweighed 37-mm, 0.8-µm pore size filter on an unweighed cellulose support pad.

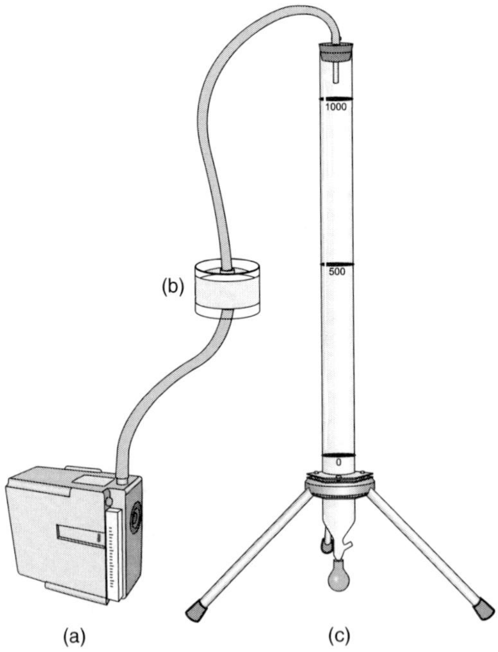

Figure 7.3 Calibration train for metal dust and fume composed of (a) a high-flow pump connected with flexible hose to (b) a 37-mm MCE filter in a three-stage cassette in-line with (c) a frictionless bubble-tube.

(iii) Conducting Monitoring

- A labeled plastic cassette is positioned with the inlet port facing downward and attached within the breathing zone, and the air sampling pump is clipped to the belt of the worker for personal monitoring. The sampling train is positioned and secured in a specific location for area monitoring.
- The pump is turned "ON" and start time is recorded.
- After specified time, the pump is turned "OFF" and stop time is recorded.
- Orifice plugs are inserted into the inlet and outlet ports of the plastic cassette. The sample is then transported to laboratory for analysis.

ANALYSIS

Metallic dusts and fumes are examples of specific particulate air contaminants that require an analytical method that will not only measure the amount of particulate collected, but will also identify specific metallic analytes (e.g., Pb, Cd, As).

Analysis of the sample typically involves either atomic absorption spectroscopy (AA) or inductively coupled plasma emission spectroscopy (ICP). Analysis of metallic particulates in the form of dust and fumes requires both detection of the specific metal or metals and measurement of the amount collected. Samples are prepared for analysis by digesting and ashing the MCE filter so that the metal air contaminant is present as a dissolved analyte. The dissolved analyte is injected into the analytical instrument for qualitative and quantitative analysis. A detailed description of these instruments is beyond the scope of this book. Nonetheless, a brief outlined summary of the components and the principles using atomic absorption spectroscopy (Figure 7.4) follows.

(i) Components

- Hollow cathode ray tube/lamp: generates a specific wavelength of light required for detection of specific metallic analyte by converting ground state atoms to an excited state
- Sample aspirator port: sample solution is automatically pumped into and nebulized in the atomic absorption spectrometer
- Burner/flame: sample chamber where temperature (2300 to 3100 K) converts metal ions in sample solution to ground state metal atoms via process called atomization
- Fuel (C_2H_2) + oxidizer (air): sustains flame in burner
- Monochromator lens: adjusted to wavelength absorbed by analyte
- Detector: photomultiplier tube that measures the intensity of light transmitted through the sample (i.e., the light not absorbed by the sample)
- Readout: receives electronic signal from detector and presents data as absorbance (Abs); commonly interfaced with a computer

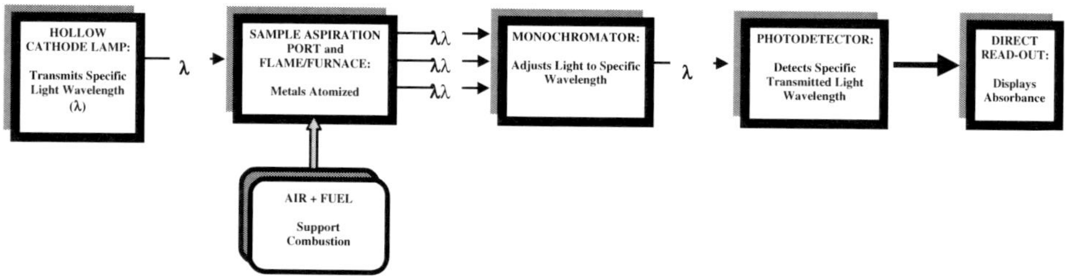

Figure 7.4 Components of an atomic absorption spectrometer.

(ii) Principle of Operation

- Specific hollow cathode ray tube/lamp that generates specific wavelength of light absorbed by the metal atoms being analyzed is selected.
- Sample solution is aspirated into the flame where specific wavelength of light from the lamp is directed.
- The high-temperature flame converts metal ions (analyte) into ground state metal atoms.
- Absorption of a specific wavelength of light from the cathode ray tube by specific metal atoms that absorb that wavelength of light causes electrons in the metal atoms to elevate to an excited state then return to ground state. A sample blank (reagent solution without metal analyte) would permit 100% transmittance and no absorbance of light from the cathode ray tube. If present, the specific metal atoms will absorb the wavelength of light linearly with their concentration (Beer's law).
- Light not absorbed by the atoms exits the sample chamber/flame and is transmitted to the detector via a monochromator adjusted to a wavelength which is absorbed by the analyte.
- The detector measures the intensity of light transmitted (i.e., not absorbed by metal atoms) and generates an electrical signal to the readout.
- The detected and measured transmitted light is converted to absorbance and displayed on a direct readout.

(iii) Determination of Concentration of Metal Dust or Fume

- Prepare a gradient of concentrations (e.g., 0 to 1000 μg/ml) of standard solutions ($Metal_{std}$) containing metal of interest (e.g., Pb).
- Aspirate aliquots of the standards into atomic absorption spectrometer and record absorbance for each known concentration.
- Plot absorbance vs. concentration of metal ($Metal_{std}$) in units of micrograms per milliliter (μg/ml) (Figure 7.5).
- Compare absorbance reading for the analyzed sample (unknown concentration) to the readings plotted for the analyzed standards (known concentrations). Extrapolate from the standard curve the concentration of metal (μg/ml) in the sample solution that was aspirated into and detected and measured by the atomic absorption spectrometer. Alternatively, calculate the linear relationship using the equation $y = mx + b$ and determine concentration.

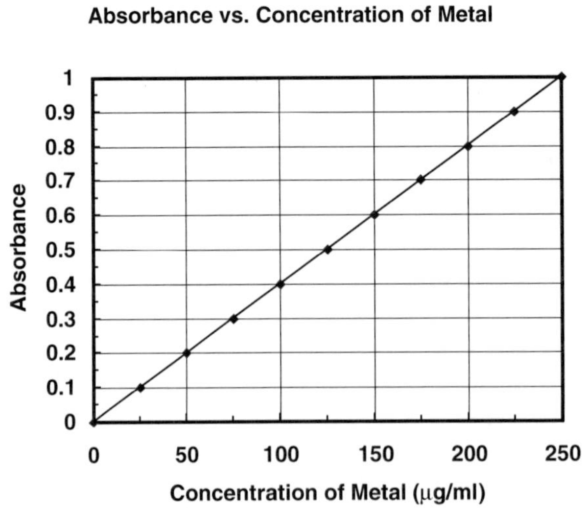

Figure 7.5 Standard curve for atomic absorption analysis of known concentrations of metal in solution.

- Calculate the amount (weight) of the metal (e.g., Pb) in the analyzed sample by multiplying the concentration of metal in the sample solution (C_{samp} [μg/ml]), based on the extrapolation from the standard curve, by the final volume of sample solution (V_{samp} [ml]) used to dissolve the MCE filter and metal in preparation for and prior to analysis.

$$\text{Wt. Metal (mg)} = \text{V}_{\text{samp}}\ (\text{ml}) \times C_{samp}\ (\text{Metal }[\mu g\,] / \text{Solution [ml]}) \times \frac{1\ \text{mg}}{10^3\ \mu\text{g}} \quad (7.3)$$

- Determine sampled volume of air by multiplying sampling time (T [min]) by average sampling flow rate (Q [l/min]) and convert liters (l) to cubic meter (m^3).

$$\text{Air Volume } (m^3) = T\ (\text{min}) \times Q_{avg}\ (\text{l / min}) \times \frac{1\ m^3}{10^3 l} \quad (7.4)$$

- Determine the concentration of metal dust or fume by dividing the weight of metal (mg) by the sampled volume of air (m^3) and express concentration in units of milligrams of metal per cubic meter of air (mg/m^3).

$$Metal\ (\text{mg / m}^3) = \frac{\text{Wt. Metal (mg)}}{\text{Air Volume (m}^3)} \quad (7.5)$$

- Note: The adjusted concentration (C) can be calculated by subtracting the weight of metal measured on a blank MCE filter (B) from the sample MCE filter (S); dividing by air volume sampled (V); and also correcting for sampling efficiency (E), which is equal to ≤ 1 depending if 100% or less efficiency is attained.

$$C = \frac{S - B}{E\,V} \quad (7.6)$$

UNIT 7 EXERCISE

OVERVIEW

The exercise will provide the fundamental concepts for conducting integrated monitoring for a metal dust or fume using a filter medium. In addition, a common and related analytical method using atomic absorption spectroscopy is introduced. The exercise will focus on sampling and analysis of lead dust or fume. The material and methods are based on a modification of the National Institute for Occupational Safety and Health (NIOSH) Method #7082. Place a check mark (✔) in the open box (☐) when you have obtained applicable material and completed steps for the sampling and analytical methods.

MATERIAL

1. Calibration of Air Sampling Pump

☐ Multi- or high-flow air sampling pump
☐ 37-mm, 0.8-μm pore size MCE filter with cellulose support pad in assembled three-stage plastic cassette

- ☐ Two 2- to 3-ft lengths of 1/4-in. i.d. flexible hose
- ☐ Manual or electronic frictionless bubble-tube calibrator
- ☐ Soap solution
- ☐ Stopwatch (if manual calibrator is used)
- ☐ Calibration data form (Figure 7.6)

2. Sampling for Metal Dust or Fume

- ☐ 37-mm, 0.8-μm pore size MCE filter with cellulose support pad in assembled 3-stage plastic cassette
- ☐ Orifice plugs
- ☐ Cellulose cassette bands or tape to secure stages of the cassette (optional)
- ☐ Multi- or high-flow air sampling pump
- ☐ One 2- to 3-ft length of 1/4-in. i.d. flexible hose
- ☐ Field monitoring data form (Figure 7.7)

3. Analysis of Metal Analyte

- ☐ Atomic absorption spectrometer with air-acetylene burner and hollow cathode ray lamp for lead
- ☐ Regulators and separate cylinders of purified compressed acetylene and air
- ☐ 50- or 125-ml beakers with watchglass covers
- ☐ 25- and 100-ml volumetric flasks
- ☐ 50- to 500-μl pipettes
- ☐ Hot plate with surface temperature 140°C
- ☐ Distilled/deionized H_2O
- ☐ Concentrated nitric acid (concentrated HNO_3)
- ☐ Nitric acid (HNO_3 10% w/v); carefully add 100 ml concentrated HNO_3 to 500 ml distilled/deionized H_2O and dilute to 1 l volume
- ☐ Hydrogen peroxide (H_2O_2 30% w/v)
- ☐ Commercially available standard solutions for lead, or mix elemental lead with hydrochloric acid (HCl) to achieve concentrations of Pb covering range from 0 to 500 μg/ml
- ☐ Laboratory analysis data form (Figure 7.8)

METHOD

1. Pre-sampling Calibration of Air Sampling Pump

- ☐ Remove pump from charger, turn "ON," and allow it to operate for 5 min prior to calibrating.
- ☐ Complete a calibration form, remembering to record name, location, date, air temperature, air pressure, pump manufacturer and model, calibration apparatus, times for three trials, and average flow rate for pre- and post-sampling calibrations.
- ☐ Obtain a representative sampling medium for use during calibration of the air sampling pump. This should consist of a 37-mm, 0.8-μm pore size MCE filter and cellulose support pad in assembled three-stage plastic cassette.
- ☐ Insert cassette adapter into one end of a 2- to 3-ft length of flexible hose.
- ☐ Connect the other end of the flexible hose to the air sampling pump.
- ☐ Connect the adapter end of the flexible hose to the outlet orifice located on the outlet stage or the support pad side of the cassette used as the representative sampling medium.
- ☐ Insert another cassette adapter into one end of another 2- to 3-ft length of flexible hose.
- ☐ Connect the adapter end of the second flexible hose to the inlet orifice located on the inlet stage or filter side of the cassette used as the representative sampling medium.
- ☐ Connect the remaining open end of the flexible hose to a manual or automatic frictionless bubble-tube calibration device. This constitutes the calibration train.
- ☐ Aspirate the bubble solution into the calibration device to form a soap film (bubble).

Calibration Data Form:
Low-Flow, High-Flow, and Multi-Flow Air Sampling Pumps

Name and Location of Calibration: ______________________

Calibration Conducted By: ______________________ **Date Calibration Conducted:** ________

Calibration Instruments (Type/Manufacturer/Model):______________________

Air Sampling Pump
(Type/Manufacturer/Model):______________________

Collection Medium In-Line: ______________ **Calibrator Volume:** ________cc

Air Temperature: ____°C **Air Pressure:** ____mm Hg **Relative Humidity:** ____%

Pump No.	Pre-Sampling Calibration Flow Rate (Q_{pre})					Post-Sampling Calibration Flow Rate (Q_{post})					Average Flow Rate
	Time (sec)				Q_{pre}	Time (sec)				Q_{post}	Q_{avg}
	T_1	T_2	T_3	T_{avg}	L/min	T_1	T_2	T_3	T_{avg}	L/min	L/min

Calibration Notes:

Figure 7.6 Calibration data form for high-flow air sampling pump.

Field Monitoring Data Form:
Integrated Monitoring for Metal Dust and Fume

Facility Name and Location: ______________________ **Contaminant Sampled:** ☐ **Metal Dust** ________
☐ **Metal Fume** ________

Monitoring Conducted By: ______________________

Collection Medium: ____________ **Date Monitoring Conducted:** ____________

Method(s)(Source/Number/Name): ______________________

Monitoring Instruments (Type/Manufacturer/Model): ______________________

Air Temperature: ____°C **Air Pressure:** ____mm Hg **Relative Humidity:** ____%

Identification of Personnel or Area	Field Sample	Pump No.	Pre-Sample Calibration Flow Rate	Post-Sample Calibration Flow Rate	Average Flow Rate (Q)	Sample Start-Time	Sample Stop-Time	Total Sample Time (T)	Volume Sampled (Vol)

Field Notes:

Figure 7.7 Field monitoring data form for metal dust and fume.

Laboratory Analysis Data Form:
Atomic Absorption Spectrophotometry Analysis of Metal Analytes

Monitored Facility Name and Location: ______________________ **Contaminant Sampled:** __________

Monitoring Conducted By: ______________________ **Date Monitoring Conducted:** __________

Method(s)(Source/Number/Name): ______________________

Laboratory Facility Name and Location: ______________________ **Contaminant Analyzed:** __________

Analysis Conducted By: ______________________ **Date Analysis Conducted:** __________

Atomic Absorption Spectrophotometer (Type/Manufacturer/Model): ______________________

Hollow Cathode Lamp: ______ **Wavelength:** ______ **Fuel Gas:** ______ **Oxidant Gas:** ______

Analytical Method: ______________________

Air Temperature: ____°C **Air Pressure:** ____mm Hg **Relative Humidity:** ____%

Laboratory Sample Identification No.	Field Sample Identification No.	Volume Prepared Sample (ml)	Volume Sample Aliquot (ml)	Dilution Factor	Absorbance	Measured Amount of Analyte (mg)	Volume Air Sampled (m^3)	Concentration of Air Contaminant (mg/m^3)

Laboratory Notes:

Figure 7.8 Laboratory analysis data form for atomic absorption analysis of metal dust and fume.

- ☐ Record the time (T [sec]) it takes the soap film (bubble) to traverse a volume of 500 to 1000 ml.
- ☐ If a manual bubble-tube is used, calculate the flow rate (Q [l/min]). Otherwise read the flow rate directly from the electronic bubble-tube calibrator.
- ☐ If the flow rate is not 2 l/min ±5%, adjust the flow control on the pump and repeat calibration procedure until desired flow rate is achieved.
- ☐ Once the desired flow rate is achieved (in this example 2 l/min ±5%), repeat the calibration check two to three times and calculate the average pre-sampling flow rate.
- ☐ Disconnect the pump from the remainder of the calibration apparatus and turn "OFF." The calibrated air sampling pump is ready for use.

2. Prepare Sampling Media

- ☐ Using a forceps, place a cellulose support pad into the outlet stage of the three-stage cassette.
- ☐ Using a forceps, gently remove an unweighed MCE filter from the package and place it on the support pad.
- ☐ Place the center ring of the three-stage cassette on the filter.
- ☐ Place the inlet stage onto the second stage.
- ☐ Press the three stages of the cassette together by using your palm and a flat rigid plastic or metal plate to evenly apply pressure.
- ☐ Place the orifice plugs into the ports in the inlet and outlet stages of the cassette.
- ☐ Place a wet cellulose band around the periphery of the cassette and allow to dry and shrink. Tape can be used instead (optional step).
- ☐ Label the cassette with a sample identification number.

3. Air Sampling for Metal Dust

- ☐ Get approval to conduct monitoring in an area where airborne levels of lead dust and fume are likely to be detectable and measurable, and perform monitoring accordingly. Alternatively, conduct a simulated monitoring exercise and obtain a sample spiked with lead from your instructor for later analysis.
- ☐ Complete a field data form, remembering to record name, location, person or area sampled, date, air temperature, air pressure, relative humidity, analyte (e.g., lead dust), air sampling pump manufacturer and model, pump identification, sampling medium, sample identification, flow rate, start and stop times for sampling, sampling duration, and air volume sampled.
- ☐ Turn "ON" pump and let operate for 5 min prior to connecting flexible hose and sample collection medium.
- ☐ Connect flexible hose to pump.
- ☐ Place and secure pump in area where sample will be collected.
- ☐ Insert a cassette adapter into other end of flexible hose.
- ☐ Remove orifice plugs from a cassette and attach cassette to adaptor positioned in flexible hose that is connected to pump.
- ☐ Record the sample number and start time and allow sample to run for at least 2 h depending on conditions.
- ☐ After an acceptable sampling period has elapsed, turn "OFF" pump, record stop time, and calculate sample time (T [min]).
- ☐ Remove cassette from the sampling train and insert the orifice plugs in the inlet and outlet stages. Transport sample to laboratory.
- ☐ Carefully wipe the outside of the cassette to remove any surface particulate that may have adhered to the surface.

4. Post-sampling Calibration of Air Sampling Pump

- ☐ Repeat steps summarized in Method #1, except do not adjust the flow rate of the pump.

☐ Postsampling flow rate (Q_{post}) should be within ±5% of the pre-sampling flow rate (Q_{pre}).
☐ Determine average sampling flow rate (Q_{avg}) by averaging the pre- and post-sampling flow rates and express average flow rate as liters per minute.

5. Analysis of Sample for Metal Dust or Fume

☐ Prepare or obtain standard solutions of known concentrations of lead. Aspirate aliquots into the atomic absorption spectrometer, measure absorbance, and plot standard curve.
☐ Remove the cellulose band or tape from around the periphery of the cassette if present.
☐ Disassemble the cassette by carefully removing the middle and inlet stages to expose the filter.
☐ Using a probe, gently push through the outlet stage orifice so that the support pad and filter are elevated and accessible for grasping with a forceps.
☐ Using a forceps, gently remove the filter and place it in a clean beaker.
☐ Add 3 ml concentrated HNO_3 and 1 ml 30% H_2O_2, cover with watchglass, heat on hot plate at 140°C, and evaporate until approximately 0.5 ml remains. Repeat two more times.
☐ Heat on hot plate at 140°C until a white ash appears.
☐ When sample is dry, rinse the watchglass and internal walls of beaker using 3 to 5 mL 10% HNO_3.
☐ Evaporate the solution to dryness, allow beaker to cool, and add 1 ml concentrated HNO_3 to dissolve residue.
☐ Transfer the solution to a 10-ml volumetric flask and dilute to 10 ml volume with distilled/deionized H_2O.
☐ Adjust atomic absorption spectrometer according to specifications established by the manufacturer and the method using an air-acetylene flame and wavelength 283.3 nm.
☐ Run sample and blank, measure absorbance, and compare against a standard curve of known concentrations vs. absorbance.
☐ Calculate the concentration of airborne lead dust in units of milligrams per cubic meters.
☐ Complete a laboratory analysis data form, remembering to record name, location, date, air temperature, air pressure, analytes (metal [e.g., Pb]), atomic absorption spectrometer manufacturer and model, sample identification, air volume sampled, and concentration.

UNIT 7 EXAMPLES

Example 7.1

A high-flow air sampling pump was calibrated using a manual frictionless bubble-tube. What was the flow rate (Q) in liters per minute (l/min) if it took an average (n = 3 trials) of 13.5 sec (T) for the bubble to traverse a volume (V) of 500 cm³?

Solution 7.1

$$Q\ (\mathrm{l/min}) = \frac{V\ (cm^3)}{T\ (\mathrm{sec})} \times \frac{60\ \mathrm{sec}}{\mathrm{min}} \times \frac{1\ \mathrm{l}}{10^3\ cm^3}$$

$$= \frac{500\ cm^3}{13.5\ \mathrm{sec}} \times \frac{60\ \mathrm{sec}}{\mathrm{min}} \times \frac{1\ \mathrm{l}}{10^3\ cm^3}$$

$$= 2.2\ \mathrm{l/min}$$

Example 7.2

A sample was collected for 2 h (T [min]) using a high-flow sampling pump at an average flow rate (Q) of 2.2 l/min. What was the volume (V [m³]) of air sampled by the pump (Note: 2 h = 120 min)?

Solution 7.2

$$\text{V } (m^3) = Q \text{ (l / min)} \times T \text{ (min)} \times \frac{1\, m^3}{10^3 l}$$

$$= 2.2 \text{ l / min} \times 120 \text{ min} \quad \times \frac{1\, m^3}{10^3 l}$$

$$= 0.264\ m^3$$

Example 7.3

A 0.264 m³ volume of air was sampled for lead dust using a high-flow air sampling pump connected to a three-stage plastic cassette with a 37-mm, 0.8-µm pore size MCE filter and support pad medium. The filter was prepared for analysis by digesting and redissolving in 10 ml 10% HNO_3. Laboratory analysis of the solution using atomic absorption spectroscopy and comparison to a standard curve indicated 23 mg Pb/ml solution. What was the concentration (mg/m³) of lead dust in the air during the sampling period?

Solution 7.3

$$Pb \text{ (mg / m}^3) = v \text{ (ml)} \times \frac{\text{Pb / ml (mg)}}{\text{Air Volume (m}^3)}$$

$$= 10 \text{ ml} \times \frac{0.023 \text{ mg Pb / ml}}{0.264 \text{ m}^3}$$

$$= 0.87 \text{ mg / m}^3$$

UNIT 7 CASE STUDY: SAMPLING AND ANALYSIS OF AIRBORNE METAL DUST

OBJECTIVES

Upon completion of Unit 7, including sufficient reading and studying of this and related reference material, and review of the short case study and hypothetical industrial hygiene sampling and analytical data presented below, learners will be able to:

- Calculate the concentrations and time-weighted averages (TWAs).
- Interpret the data and results.
- Concisely summarize the information in a concise report.
- Retain the general scientific and technical principles and concepts related to the topic.

INSTRUCTIONS

After reading the short case study, locate and read the applicable NIOSH method for sampling and analysis (http://www.cdc.gov/niosh/nmam/nmammenu.html). Next, review the data shown in the tables below and perform the necessary calculations to complete the tables (assume normal temperature pressure). Review and interpret the results, and compare calculated TWAs to applicable occupational exposure limits (e.g., Occupational Safety and Health Administration permissible exposure limits [OSHA-PELs], American Conference of Governmental Industrial Hygienists threshold limit values [ACGIH-TLVs], and NIOSH recommended exposure limits [NIOSH-RELs]). Also, prepare responses to the questions that follow the data section. Finally, prepare a fictional, but representative, report following the outline or similar format shown in Appendix B.

CASE STUDY

Fuels are stored in aboveground storage tanks on the property of a petroleum refinery. Historically, many of these tanks were painted with several coats of lead-based paint. Over time, the layers of paint began to deteriorate and peel. Consequently, a maintenance plan was established to scrape and sand off the peeling paint prior to resealing the tank surface with fresh paint. The industrial hygienists were concerned about generation of airborne paint dust containing lead. Accordingly, they conducted integrated sampling for airborne lead dust and monitored a maintenance worker and an adjacent area during a paint scraping and sanding operation. Sampling and analysis for airborne metal (lead) dust followed NIOSH Method #7105. Calibration, field, and lab data are summarized in the tables below.

DATA AND CALCULATIONS

Table 7.1 Calibration and Field Data

Name of Personnel or Area	Sample No.	Analyte	Avg. Pre-sample Q (l/min)	Avg. Post-sample Q (l/min)	Avg. Q (l/min)	Sample Start Time	Sample Stop Time	Sample Time (min)	Air Volume Sampled (l)
	1136					1500	1730		
Joe M.	1140	Lead	Pump No. 001: 2.16	Pump No. 001: 2.20		1730	2000		
	1128					2000	2300		
Tank 1 Railing	1138	Lead	Pump No. 005: 2.08	Pump No. 005: 2.02		1500	1900		
	1125					1900	2300		
Blank	1131	Lead	N/A	N/A	N/A	N/A	N/A	N/A	—

N/A = Not applicable.

Table 7.2 Lab Standard Curve Data for Atomic Absorption Spectrometer

Abs	0	0.1	0.2	0.3	0.4	0.5	0.6	0.7	0.8	0.9	1.0
Lead Std. (μg/ml)	0	0.25	0.50	0.75	1.00	1.25	1.50	1.75	2.00	2.25	2.50

Table 7.3 Lab Data

Name of Personnel or Area	Sample No.	Analyte	Absorbance	Conc. Analyte (μg/ml)	Volume Acid to Dissolve Filter (ml)	Mass of Analyte (mg)	Air Volume Sampled (m^3)	Conc. Lead Dust (mg/m^3)
	1136		0.70		10			
Joe M.	1140	Lead	0.50					
	1128		0.80					
Tank 1	1138	Lead	0.40		10			
Railing	1125		0.30					
Blank	1131	Lead	<LOD	N/A	10	<LOD	0	

LOD = Limit of detection.
N/A = Not applicable.

Table 7.4 Calculated Time-Weighted Averages

Name of Personnel or Area	Sample No.	Analyte	8-h TWA Lead Dust (mg/m^3)
	1136		
Joe M.	1140	Lead Dust	
	1128		
Tank 1	1138	Lead Dust	
Railing	1125		
Blank	1131	Lead Dust	

QUESTIONS

- Were the average pre- and post-sampling flow rates within ±5% of each other?
- Was the total sample volume collected within the min-max range specified in the applicable NIOSH method?
- What is the minimum number of field blanks that should be submitted to the laboratory, based on the number of samples collected and the applicable NIOSH method?
- What considerations or assumptions did you make when calculating the TWAs? For example, did you assume the workers took breaks (e.g., lunch) in areas outside the immediate area? Would you continue monitoring while workers are on breaks, even if outside the potential exposure area? If not, what if the actual sample collection period was less than 8 h (480 min)?
- What is the estimated limit of detection (LOD) for the analyte as per the applicable NIOSH method?
- How did the calculated TWAs compare to applicable occupational exposure limits, such as the OSHA-PEL, ACGIH-TLV, and NIOSH-REL?
- What was the basis for the established occupational exposure limits? (Learners and practitioners alike are strongly encouraged to obtain and read applicable information found in resources such as the NIOSH *Criteria Documents* and ACGIH *Documentation of the TLVs and BEIs*.)
- Are there other NIOSH or OSHA air sampling and analytical methods for the same agent? If so, what are the reference numbers and what are the differences relative to sample collection (e.g., medium used) and analysis (e.g., analytical instrument used)?

REPORT

Prepare a concise report (see Appendix B for outline of sample format) that includes typed text and tables (and figures if you desire). Your report needs to include a section on calibration (how you calibrate pump with representative media), sample collection (instruments, media, and

NIOSH method used), and analysis (instrument and NIOSH method used). In other words, write your short field report so the reader knows what was done to prepare for sampling and analysis for airborne metal dust. Also, summarize, interpret, and discuss the results; state your conclusions; and make applicable recommendations. Remember to add a bibliography at the end of the report citing any applicable book, journal, and Internet references that you consult. Finally, append a copy of the applicable NIOSH methods.

UNIT 8

Evaluation of Airborne Particulate: Instantaneous Area Sampling Using a Direct-Reading Aerosol Meter

LEARNING OBJECTIVES

At the completion of Unit 8, including sufficient reading and studying of this and related reference material, learners will be able to correctly:

- Name and identify a common instrument used for conducting instantaneous or real-time area sampling for airborne particulates.
- Summarize the principles of sampling using a direct-reading instrument for airborne particulates.
- Conduct instantaneous monitoring of airborne particulates using a direct-reading aerosol meter.
- Record all applicable sampling data using a field sampling data form.

OVERVIEW

In Units 4–7, instrumentation and methods for conducting integrated or continuous personal and area sampling and analysis of classes of airborne particulates were summarized. Instrumentation and methods also are available for conducting instantaneous or real-time area monitoring of generic total and respirable particulates, as well as more specific airborne particulates such as fibers. The instrumentation typically involves battery-operated electronic meters that will actively pump or blow air into the device where airborne particulates are detected and measured. Passive devices without an active flow mechanism also are available. The concentration of airborne particulate is indicated on a direct readout display (Figure 8.1).

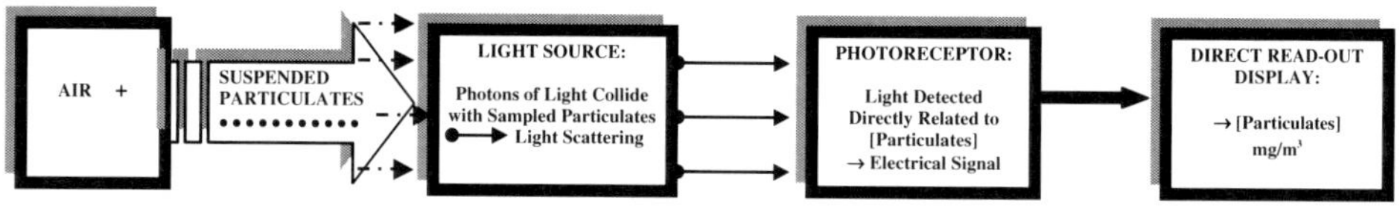

Figure 8.1 Schematic of light scattering aerosol meter.

Direct-reading instruments for monitoring airborne particulates are generally less accurate, but more convenient than integrated air monitoring techniques which require time for laboratory analysis. Their speed of operation allows rapidly changing conditions to be assessed at the specific time of sampling.

The method of obtaining samples will depend on the purpose of the monitoring. If the purpose is to determine if limits for personal exposures are being exceeded, the inlet of the monitor must be held in the worker's breathing zone for the time required for an adequate sample to be collected. More frequently, the monitor is placed in a stationary position 4 to 6 ft above the working surface to collect area samples. Area samples also are collected using handheld devices that permit movement to various locations while simultaneously sampling for airborne aerosols. Instantaneous area monitoring is important to determine the concentration of particulates within a given area, the sources of emission, the effectiveness of ventilation controls, and the peak concentrations of particulates during a process cycle.

SAMPLING

Several types of direct-reading aerosol meters are available. The devices differ based on application and method of detection and measurement. Two major principles of particulate detection and measurement are light scattering and electrical precipitation and oscillation. In addition, based on application, some devices differentiate between spherical and fibrous particulates and respirable fraction and total particulate. Air is sampled via a pumping or blowing mechanism that transports airborne particulate into the instrument.

Most applications of the instruments are for either stationary or mobile area sampling. During stationary sampling, some devices can be integrated with a data-recording device, including notebook computer, to collect samples and record real-time results over a specified period. For mobile use, the meter is carried while operating so that levels of airborne particulate can be detected and measured at various locations (Figure 8.2). Some smaller instantaneous aerosol monitors are available that connect to individuals for real-time personal sampling.

(i) Light Scattering

The light scattering properties of suspended particles can be utilized to determine aerosol size and concentration. There are a variety of light scattering instruments called photodetectors

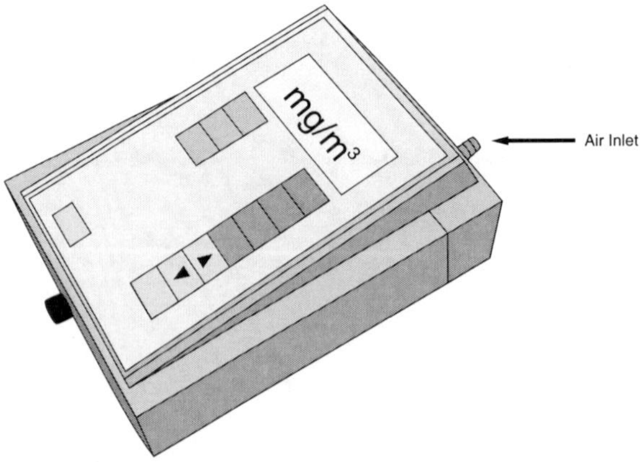

Figure 8.2 Total and respirable dust aerosol meter.

or photometers available for measuring total and respirable particulates. All such instruments incorporate a light source to illuminate the sampled particles as they enter a sensing zone. A photoelectric receptor (detector) positioned at an angle to the incident light beam detects and measures the scattered light. These elements are arranged so that the receptor collects light scattered by the particles at an angle from the incident beam. Light scattering instruments are generally divided into forward scattering photometers and angular dispersion photometers. Forward scattering photometers sense light scattered in the near forward direction from the particulate. In contrast, angular dispersion photometers detect light scattered by particles throughout an angular sensing zone. Measurements of spheroid particulates are read in units of milligrams per cubic meter (mg/m^3).

Another technology, known as condensation nuclei counters, also involves light scattering and detection. The sampled particles are first enlarged to a size that can be detected. This technology involves an instrument that collects, detects, and counts otherwise undetectable, ultrafine submicrometer (0.02 to <1 μm) particles. The sampled air containing suspended particles flows into a heated chamber where the particles mix with isopropanol vapor. The particles and vapor pass into a condenser where they cool and the alcohol vapor condenses on the particles forming larger (>1 μm), detectable, and measurable particles. Laser optics focus light on the enlarged particles which, in turn, causes light scattering. The scattered light is detected by a photodiode and converted into electrical pulses that are counted in direct relationship to the concentration of particles.

Airborne concentrations of fibrous particles, such as asbestos fibers, also can be assessed via an instantaneous monitoring device that uses light scattering technology. Sampling can involve the use of a meter that measures the airborne concentration of fibrous materials with a length-to-diameter ratio of 3:1. One monitoring device uses a helium-neon laser and electro-optical sensors that detect fiber oscillations as the aerosol passes through a rapidly oscillating high-intensity field. Fiber concentration is measured based on variation of intensity of scattered light detected by a photoreceptor (detector). The fiber count is digitally displayed in fibers per cubic centimeter (f/cm^3).

(ii) Piezoelectric Resonant Oscillation

Another mechanism for detection and measurement of airborne particulates involves collection via electrostatic attraction followed by detection and measurement based on resonant oscillation. Piezoelectric monitoring instruments collect spherical aerosol particles on the surface of a piezoelectric quartz crystal via electrostatic precipitation and mechanical impaction. As the piezoelectric crystal is oscillated, particulates depositing on the surface of the crystal change its resonant frequency. The resonant frequency of the crystal changes inversely as the mass deposited on its surface is increased. The net change in the resonant frequency of the piezoelectric quartz crystal during the sampling period is determined and converted to mass concentration. The concentration of particulate is then displayed as milligrams per cubic meter.

UNIT 8 EXERCISE

OVERVIEW

The exercise will provide the fundamental concepts for conducting instantaneous monitoring of airborne particulates using an aerosol survey meter. Place a check mark (✔) in the open box (☐) when you have obtained applicable material and completed the steps for the sampling method.

MATERIAL

1. Calibration of Aerosol Survey Meter

- ☐ Factory or manufacturer calibration is typically adequate. Instruments typically require assuring the display is adjusted to zero using a zero reference source prior to each use.

2. Sampling for Airborne Particulates

- ☐ Aerosol survey meter
- ☐ Tape measure or ruler
- ☐ Field sampling data form (Figure 8.3)

METHOD

1. Pre- and Post-sampling Calibration of Aerosol Survey Meter

- ☐ The instruments are typically factory calibrated by the manufacturer. The meter must be returned to the manufacturer or the equivalent for periodic adjustments and recalibration. Follow the manufacturer's instructions for operating the instrument.

2. Sampling for Airborne Particulate

- ☐ Get approval to conduct monitoring in an area where airborne levels of particulate are likely to be detectable and measurable and perform monitoring accordingly. Alternatively, conduct a simulated monitoring exercise.
- ☐ Turn "ON" and zero the aerosol survey meter.
- ☐ Hold the aerosol meter 4 to 6 ft above the floor or working surface and slowly scan the area.
- ☐ Read and record the concentration of particulate displayed on the direct readout.
- ☐ Measure the dimensions of the area, and read and record measurements at various locations.
- ☐ Complete a field sampling data form, remembering to record your name, facility and location sampled, date, air temperature, air pressure, relative humidity, and aerosol meter manufacturer and model.

Field Monitoring Data Form:
Real-Time Monitoring for Aerosols
Using An Aerosol Meter

Facility Name and Location: ______________________________

Monitoring Conducted By: ______________________________

Date Monitoring Conducted: ____________

Monitoring Instrument
(Type/Manufacturer/Model): ______________________________

Pre-Sampling Reference Check or Zero: ______

Post-Sampling Reference Check or Zero: ______

Suspected Contaminant(s): ______________________________

Air Temperature: ____°C **Air Pressure:** ____mm Hg **Relative Humidity:** ____%

Identification of Area	Concentration (mg/m^3)	
	Total	Respirable

Field Notes:

Figure 8.3 Field sampling data form for instantaneous monitoring of total or respirable dust.

UNIT 9

Evaluation of Airborne Organic Gases and Vapors: Integrated Personal and Area Monitoring Using an Air Sampling Pump with a Solid Adsorbent Medium

LEARNING OBJECTIVES

At the completion of Unit 9, including sufficient reading and studying of this and related reference material, learners will be able to correctly:

- Name, identify, and assemble the components of a common sampling train, including the specific sampling medium, used for conducting integrated sampling of airborne organic vapors using solid adsorbents.
- Name, identify, and assemble the applicable calibration train that uses a primary standard for measuring flow rate of a multi- or low-flow air sampling pump.
- Calibrate a multi- or low-flow air sampling pump, with the applicable representative sampling medium in-line, using a manual or electronic frictionless bubble-tube.
- Summarize the principles of sample collection of airborne organic vapors using solid adsorbent media.
- Conduct integrated sampling of airborne organic vapors using solid adsorbent media.
- Prepare samples for analysis of collected vapor using gas chromatography.
- Name and identify the components of a gas chromatograph.
- Name, identify, and assemble the components necessary for analysis of samples for organic vapors using a gas chromatograph.
- Calibrate a gas chromatograph and establish a standard curve using known concentrations of an organic analyte.
- Summarize the principles of sample analysis of solid adsorbents used for collection of airborne organic vapors.
- Conduct analysis of a sample for vapor using gas chromatography.
- Perform applicable calculations and conversions related to pre- and post-sampling flow rate of an air sampling pump, average flow rate of an air sampling pump, sampling time, sampled air volume, weight of organic analyte collected, and concentration of organic vapor in units of parts per million (ppm).
- Record all applicable calibration, sampling, and analytical data using calibration, field sampling, and laboratory analysis data forms.

OVERVIEW

Nonparticulate air contaminants include gases and vapors for which occupational exposure limits and specific sampling and analytical methods have been established. Airborne gases and vapors can be separated from a contaminated airstream via adsorption to a solid medium following contact. Accordingly, some integrated sampling methods for gases and vapors involve sorptive techniques using a solid medium, such as activated carbon or silica gel, as adsorbents. Adsorption refers to the immediate bonding of a gas or vapor to the surface of a solid under normal conditions of temperature and pressure. Adsorption specifically refers to bonding at the surface interface where contact occurs between the respective surfaces of the solid adsorbent and the flowing gas or vapor molecules (Figure 9.1).

Solid adsorbent media are characterized as small, porous spheroid particles. This characteristic, in turn, provides high specific surface area (i.e., square millimeters of surface area/gram of medium) for contact between airborne gases and vapors and the surface of the adsorbent. In turn, adsorption efficiency is directly related to the amount of specific surface area that is exposed. Other factors that affect collection efficiency include sample flow rate, concentration of the gas or vapor, temperature, and humidity. If flow rates are too high, not enough time is allowed for the contaminant to interact or bond with the adsorbent. The lowest flow rate that will give a sample of sufficient size for analysis should be utilized, but flow rates are rarely less than 10 cm^3/min. To compensate for high concentrations of gases or vapors, sampling time and flow rates can be reduced or a larger sorbent tube containing more adsorptive media can be utilized. Both elevated temperature and humidity can increase the rate at which the surface of adsorbent media becomes saturated.

SOLID ADSORBENTS FOR ORGANIC GASES AND VAPORS

The most common solid adsorbent media are activated charcoal and silica gel. Activated charcoal implies that the medium was activated by heating a carbonaceous medium to several hundred degrees Celsius. A common form is activated coconut shell charcoal. Activated charcoal is used for the collection of nonpolar organic vapors with boiling points greater than 100°C. Although the carbon medium is not hygroscopic, high moisture levels in the monitored atmosphere can significantly reduce collection efficiency. Adjustments in sampling strategy are necessary when relative humidity exceeds 80%. Silica gel is generally used to collect more polar substances that do not efficiently adsorb to activated carbon. Silica gel is synthesized from a silicate salt and sulfuric acid and is used to collect polar organics. These include compounds containing hydroxyl groups, halogenated hydrocarbons, and inorganic compounds. Ranked in order of decreased sorption efficiency, silica gel will collect water, alcohols, aldehydes, ketones, esters, aromatics, unsaturated aliphatics, and saturated aliphatics. The silica medium is very hygroscopic, and, accordingly, collection efficiency decreases under conditions of high relative humidity, and also when levels exceed 80%.

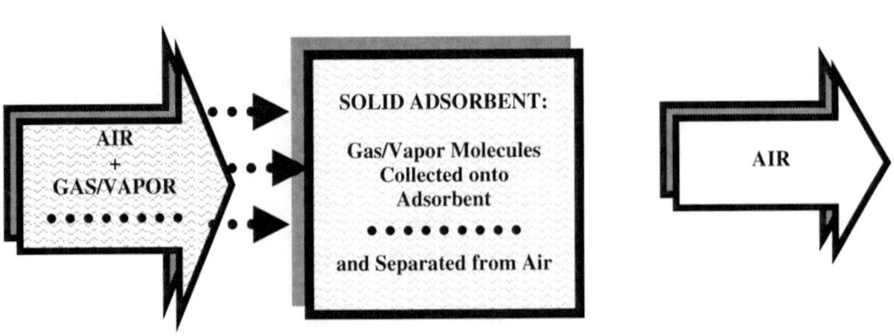

Figure 9.1 Separation of organic vapor samples from air via adsorption on a solid medium.

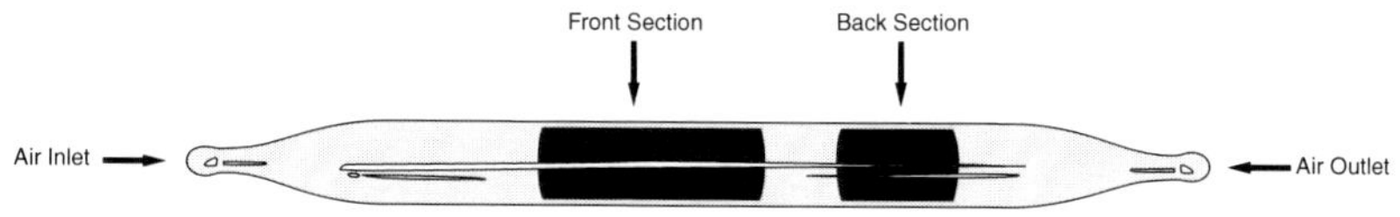

Figure 9.2 Standard glass tube filled with a solid adsorbent medium.

The media are contained within flame-sealed glass tubes tapered at each end (Figure 9.2). A glass fiber plug is wedged in the inlet end of the tube and a 3-mm urethane foam plug in the outlet end. The adsorbent medium is the tube is divided into two sections, referred to as the front and back sections. The two sections are separated typically by a 2-mm urethane foam plug. A standard charcoal tube has an internal diameter (i.d.) of 4 mm, an outer diameter of 6 mm, and a length of 70 mm. Each tube contains two beds of 20/40 mesh coconut carbon. The main or front section of the tube contains 100 mg of activated carbon and the back section contains 50 mg of carbon. A standard silica gel tube contains two sections of 20/40 mesh silica gel. The front section contains 130 mg of silica gel and the back section contains 65 mg of silica gel.

Larger tubes containing more media to provide increased specific surface area for adsorption are available for both activated charcoal and silica gel. This permits sampling at higher flow rates and collection of higher sample volumes. For example, this is one method for collecting a bulk sample of air for qualitative identification of organic compounds, as discussed in Unit 14.

An increasing number of coated and specialty adsorbents are used for highly reactive and volatile organic compounds. Examples include the commercially available polymers and resins with trade names such as Anasorb, Chromosorb, Florisil, Porapak, Tenax, and XAD adsorbents contained within sealed glass tubes.

PRECAUTION

Theoretically, when a gas or vapor is drawn into an adsorbent tube, the molecules are retained in the front section. As flow continues through the tube, it is possible for some of the gas or vapor to continue movement to the back section. When this occurs, it is defined as breakthrough. The purpose of the backup section media in the sampling tubes is to detect if contaminant breakthrough has occurred.

There are several factors that cause an agent to pass from the front to the back section of media in sampling tubes. A primary factor is saturation of the front section of an adsorbent medium. This condition can result when the concentration of gas or vapor in the air is relatively high and sampling flow rates and times are not adjusted to account for this factor. As a result, the possible bonding sites on the adsorbent medium for gas and vapor molecules are saturated which, in turn, permit the molecules to bypass the front section and flow to the back section. Another factor that can influence breakthrough is desorption of the gas or vapor from the front section of a medium and subsequent passive diffusion to the back section. This can occur following sampling, especially if the sampling tube is not capped on both ends and stored at $<4°C$ during the interim prior to analysis. A phenomenon known as channelling can also result in breakthrough of contaminant. Channelling may result when a sorbent tube is positioned horizontally during sample collection. Horizontal positioning of the tube potentially allows the adsorbent media to shift and form a channel. When contaminated air is drawn through the tube, the channel may permit gases and vapors to bypass the medium in the front section, resulting in breakthrough.

The back section of media in adsorbent tubes provides a means of checking whether breakthrough has occurred. Following sampling, the front and back sections are analyzed separately. If the amount of analyte detected and measured on the back section is greater than 10% of the amount on the front section, the validity of the sample is questionable.

SAMPLING

Prior to sampling, both ends of a glass sampling tube are carefully snapped open for air to flow through during sample collection. Integrated sampling for airborne gases and vapors involves a low- or multiflow pump set at a flow rate less than 1000 cm^3/min and commonly lower than 100 cm^3/min. The pump is connected to a flexible hose and tube holder assembly which, in turn, contains the glass tube packed with the solid adsorbent (Figure 9.3). The sampling pump creates a vacuum which causes air to be pulled into the tube and through the front, then the backup sections prior to being exhausted from an outlet port on the pump. In theory, if gases or vapors are present in the air as it passed through the column of solid medium particles, they would be immediately removed from the airstream and collected via adsorption. Subsequent to sample collection, the sampling tube is removed and plastic caps are placed on each end. Tubes are typically incubated at <4°C to minimize migration of volatile analytes during the interim between sampling and laboratory analysis. Most common analysis involves gas chromatography. A brief outlined summary of sample collection follows below.

(i) Calibration

- An air sampling pump is calibrated pre- and post-sampling to adjust or determine flow rate using a manual or electronic calibrator with a glass tube containing applicable solid sorbent in-line (Figure 9.4).
- Pre- (Q_{pre}) and post-sampling (Q_{post}) flow rates are determined by measuring the average time (T_{avg} [sec]) based on three trials ($\Sigma\ T_{1,2,3}/3$) for a bubble to traverse a specific volume (V [cm^3]) of the bubble-tube.

$$Q\,(\text{cm}^3\,/\,\text{min}) = \frac{\text{V (cm}^3\text{)}}{T_{avg}\text{(sec)}} \times \frac{60\ \text{sec}}{1\,\text{min}} \tag{9.1}$$

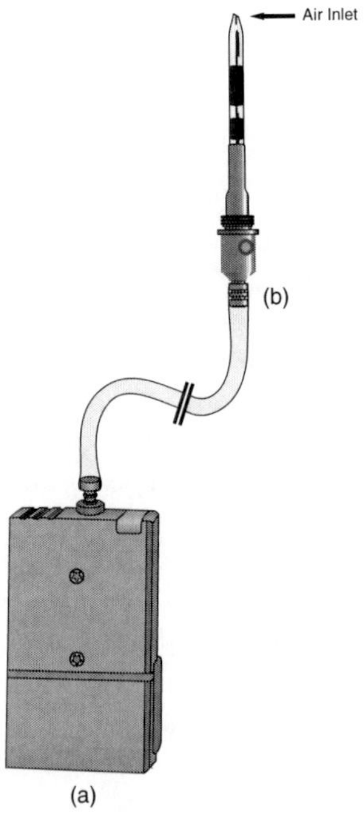

Figure 9.3 Sampling train for organic vapors composed of (a) a low-flow pump connected with flexible hose to (b) a glass tube filled with solid adsorbent medium (positioned in a tube holder without cover).

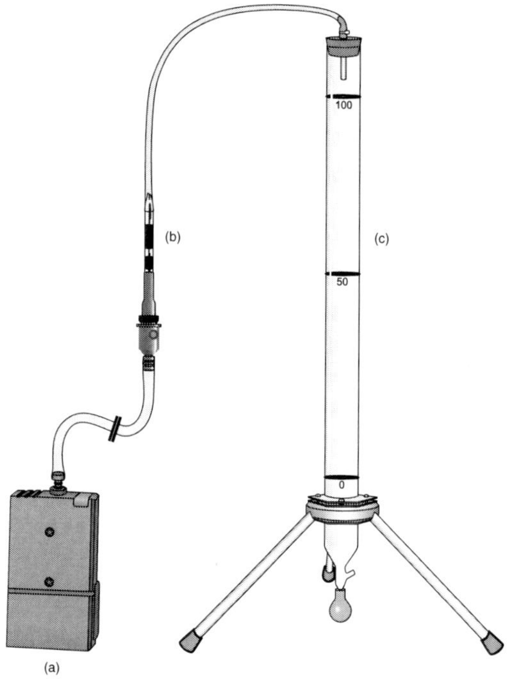

Figure 9.4 Calibration train for organic vapors composed of (a) a low-flow pump connected with flexible hose to (b) a glass tube filled with solid adsorbent medium in-line with (c) a frictionless bubble-tube.

- The average flow rate (Q_{avg}) is based on the average of pre- (Q_{pre}) and post-sampling (Q_{post}) flow rates.

$$Q_{avg} = \frac{Q_{pre} + Q_{post}}{2} \tag{9.2}$$

- Typical flow rates for sampling organic gases and vapors using solid adsorbents are 10 to 200 cm^3/min.

(ii) Preparation for Sampling

- Both ends of a labeled glass sampling tube are broken open.
- A sampling train is assembled and consists of multi- or low-flow pump with an 1/8-in. i.d. flexible hose connected to a tube holder containing sorbent tube.

(iii) Conducting Sampling

- A tube holder containing labeled sampling tube is positioned vertically and attached within the breathing zone, and the air sampling pump is secured to the belt of the worker for personal sampling. The sampling train is positioned and secured in a specific location for area sampling.
- The pump is turned "ON" and start time is recorded.
- After a specified time, the pump is turned "OFF" and stop time is recorded.
- The sample sorbent tube is capped and transported to the laboratory for analysis.

Adsorbent tubes are used when conducting active-flow, integrated air sampling (as defined in Unit 1). Alternative sampling devices called passive dosimeters (Figure 9.5) have been developed; they are continually being perfected for sampling organic gases that have been traditionally sampled

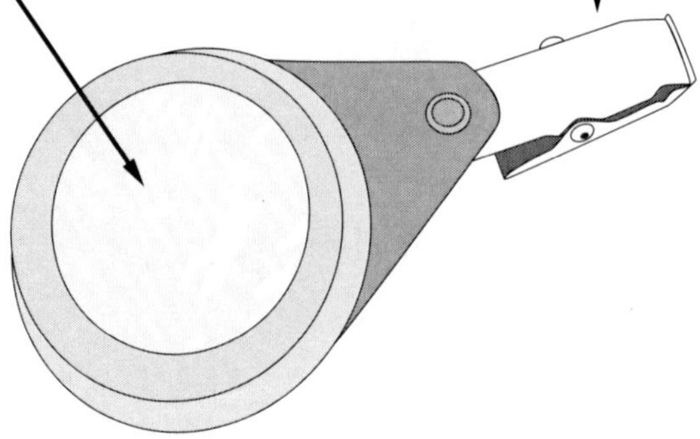

Figure 9.5 Passive dosimeter badge filled with a solid adsorbent medium.

using glass tubes containing charcoal and silica gel adsorbents. The alternative devices are referred to as "passive" dosimeters since they do not rely on active flow from an air sampling pump to collect the sample. Instead, the devices are designed to enhance the passive diffusion of airborne vapors into the dosimeter where the contaminant adsorbs to a solid sorbent, typically activated charcoal. The rate of diffusion is controlled by the dimensions of the collection device, the diffusion coefficient of the gaseous contaminant, and the distance that the contaminant must diffuse into the monitor.

ANALYSIS

Gas chromatography (GC) is the method of analysis most often utilized for contaminants collected on solid sorbent tubes. As mentioned previously, the two sections of solid adsorbent in the sampling tube are analyzed separately. Each section is removed individually and placed in separate vials. A solvent, usually carbon disulfide for charcoal tubes and methanol for silica gel tubes, is added to desorb the gas or vapor contaminant from the solid adsorbent medium. Following a short incubation period for desorption, the concentration of the contaminant can then be determined via GC (Figure 9.6). To determine the total amount of contaminant (analyte) collected for a given sample, the sum of the amounts detected and measured for each section of a tube is determined (unless the amount collected on the backup section exceeds 10% of the front section).

The media in some passive monitors, discussed briefly earlier, are also prepared for analysis and analyzed using gas chromatography in a similar manner.

A detailed description of GC is beyond the scope of this book. A brief summary of the components and principles follows.

(i) Components

- Inert carrier gas: serves as the mobile phase to transport the sample though the GC system; commonly purified N_2 or He gas.
- Sample injection port: sample is manually or automatically injected and combined with the carrier gas and is considered the mobile phase. An elevated temperature (e.g., 200 to 250°C) is maintained to vaporize the liquid sample.
- Column: a coiled glass or metal tubing containing a support medium serves as a stationary phase through which the mobile phase flows and retention and separation of analytes occurs.
- Oven: contains the column and maintains an elevated temperature (e.g., 100 to 250°C).

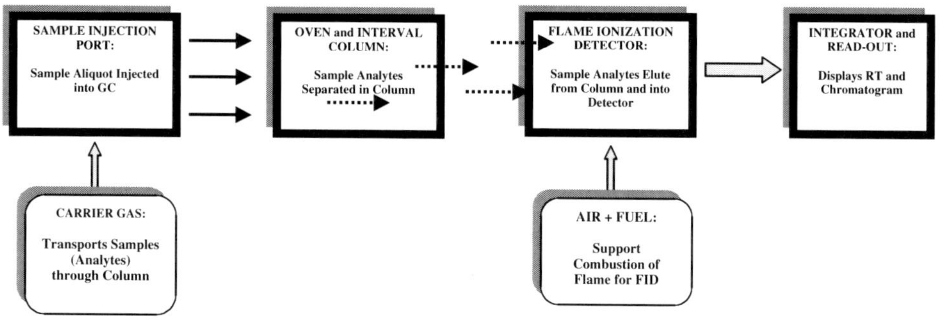

Figure 9.6 Components of a gas chromatograph.

- Detector: detects analytes that elute from the column. The most common is a flame ionization detector (FID) which measures change in ionization due to combustion of analytes. FID is used for numerous classes of organics, especially hydrocarbons.
- Fuel (H_2) + oxidizer (air): support combustion in FID.
- Integrator/recorder: receives an electrical signal from detector and presents and plots (chromatogram) data. It provides retention time (RT) and area under the curve for each analyte and is commonly interfaced with a computer.

(ii) Principle of Operation

- An aliquot of desorption solvent containing dissolved analytes is manually or automatically withdrawn from the sample vial.
- The liquid aliquot is manually or automatically injected into the injection port. The liquid solvent and dissolved analytes are immediately vaporized subsequent to injection.
- Vaporized desorption solvent and analytes are transported (carried) through the column via the carrier gas. This constitutes the mobile phase.
- The mobile phase passes through the column and the desorption solvent, and the analytes interact with the stationary phase. Each analyte will interact differently with the stationary phase and, as a result, each analyte separates (partitions) from the others.
- Separation of the mobile phase components (desorption solvent plus analytes) is based on their vapor pressure and solubility with the stationary phase medium. If the stationary phase is nonpolar, then nonpolar analytes would exhibit greater solubility with the medium and longer RT than the more polar analytes (the reverse would hold true).
- The analytes elute off the column to a detector in the order of shortest to longest RT. As the analytes are detected, the data are plotted on a recorder as curves/peaks on a graph called a chromatogram.

(iii) Determination of Concentration of Organic Gas or Vapor

- The analytes are identified by comparing the RT for the unknown sample analyte to the RT for a known standard analyte. Analytes are quantified based on the area under the curve or peak area (area ratio).
- Prepare a gradient of concentrations (e.g., 0.01 to 10 mg/ml) of organic analyte ($Organic_{std}$) containing the organic of interest (e.g., toluene).
- Manually or automatically inject aliquots of the standards into the gas chromatograph and record peak area for each known concentration.
- Plot peak area vs. concentration of organic ($Organic_{std}$) in units of milligrams per milliliter (mg/ml). Note that peak area can vary based on injection volume. Accordingly, area ratios are actually calculated based on comparison to peaks generated from analysis of an internal reference.
- Compare the peak area ratio reading for the analyzed sample (unknown concentration) to the readings plotted for the analyzed standards (known concentrations). Extrapolate from the standard

curve the concentration of organic (mg/ml) in the sample solution that was injected into and detected and measured by the gas chromatograph.

- Calculate the amount (weight) of the organic (e.g., toluene) in the analyzed sample by multiplying the concentration of organic in the sample solution (C_{samp} [mg/ml]), based on the extrapolation from the standard curve, by the final volume of sample solution (V_{samp} [ml]) used to desorb the organic from the solid adsorbent in preparation for and prior to analysis.

$$\text{Wt. Organic (mg)} = \text{v}_{\text{samp}}\ (\text{ml}) \times C_{samp}\ (\text{Organic } [mg\ ] / \text{Solution [ml]}) \tag{9.3}$$

- Determine the sampled volume of air by multiplying sampling time (T [min]) by average sampling flow rate (Q [cm^3/min]) and convert cubic centimeters (cm^3) to cubic meters (m^3).

$$\text{Air Volume } (m^3) = T\,(\text{min}) \times Q_{avg}\,(\text{cm}^3/\text{min}) \times \frac{1\ m^3}{10^6\ \text{cm}^3} \tag{9.4}$$

- Determine the concentration of organic gas or vapor by dividing the weight of the organic (mg) by the sampled volume of air (m^3) and express concentration in units of milligrams of organic gas or vapor per cubic meter of air (mg/m^3).

$$Organic\ (\text{mg}/\text{m}^3) = \frac{\text{V}_{samp}\,(ml) \times C_{samp}\,(mg/ml)}{\text{Air Volume } (\text{m}^3)} \tag{9.5}$$

- Convert milligrams per cubic meter (mg/m^3) to parts per million (ppm).

$$Organic\ (\text{ppm}) = \frac{Organic\ (mg)}{\text{Air Volume } (\text{m}^3)} \times \frac{24.45\ \text{l}/\text{g-mol}}{\text{MW } (\text{g}/\text{g-mol})} \times \frac{1\ m^3}{10^3\ l} \times \frac{1\ g}{10^3\ mg} \times 10^6 \tag{9.6}$$

- Note: The adjusted concentration (C) can be calculated by subtracting the weight of organic analyte measured on a blank sorbent tube (B) from the total weight of analyte on a sample sorbent tube (S); dividing by the air volume samples (V); and also correcting for sampling efficiency (E), which is equal to ≤1 depending if 100% or less efficiency is attained.

$$C = \frac{S - B}{E\,V} \tag{9.7}$$

UNIT 9 EXERCISE

OVERVIEW

The exercise will provide the fundamental concepts for conducting integrated sampling for an organic vapor using a solid adsorbent medium. In addition, a common and related analytical method using GC is introduced. The exercise will focus on sampling and analysis of nonpolar aromatic hydrocarbon vapors. The material and methods are based on a modification of the National Institute for Occupational Safety and Health (NIOSH) Method #1501. Place a check mark (✔) in the open box (☐) when you have obtained applicable material and completed the steps for sampling and analytical methods. Although passive dosimeters are not used in the exercise that follows, obtain several examples for observation and comparison.

MATERIAL

1. Calibration of Air Sampling Pump

- ☐ Multi- or low-flow air sampling pump
- ☐ Solid sorbent tube containing 150-mg (100/50) coconut shell charcoal
- ☐ Sampling tube end-breaker
- ☐ Sorbent sampling tube holder
- ☐ Two 2-ft lengths of 1/8-in. i.d. flexible plastic tubing
- ☐ Manual or electronic frictionless bubble-tube calibrator
- ☐ Soap solution
- ☐ Stopwatch (if manual calibrator is used)
- ☐ Calibration data form (Figure 9.7)

2. Sampling for Nonpolar Organic Vapor

- ☐ Solid sorbent tube containing 150-mg (100/50) coconut shell charcoal
- ☐ Sampling tube end-breaker
- ☐ Plastic end caps
- ☐ Multi- or low-flow air sampling pump
- ☐ Sorbent tube holder
- ☐ One 2-ft lengths of 1/8-in. i.d. flexible hose
- ☐ Field sampling data form (Figure 9.8)

3. Sample Analysis of Nonpolar Organic Analyte

- ☐ File, probe, and forceps
- ☐ Carbon disulfide (analytical grade)
- ☐ Standard toluene and xylene (or other applicable) analytes in CS_2
- ☐ GC with FID, 3-m × 2-mm glass capillary column with 10% OV-275 on 100/120 mesh Chromosorb W-AW or equivalent and appropriate compressed carrier (purified N_2 or He), oxidizer (purified air), and fuel (H_2) gases
- ☐ 2-ml GC vials with caps and septa (Teflon™/polytetrafluoroethylene [PTFE] lined)
- ☐ 1-ml pipette
- ☐ Microliter syringe and/or autosampler for GC injections
- ☐ Laboratory analysis data form (Figure 9.9)

METHOD

1. Pre-sampling Calibration of Air Sampling Pump

- ☐ Remove pump from charger, turn "ON," and allow it to operate for 5 min prior to calibrating.
- ☐ Complete a calibration form, remembering to record name, location, date, air temperature, air pressure, pump manufacturer and model, calibration apparatus, times for three trials, and average flow rate for pre- and post-sampling calibrations.
- ☐ Obtain a model sampling medium for use during calibration of the air sampling pump. This should consist of a solid sorbent tube containing 150-mg coconut shell charcoal.
- ☐ Insert the outlet end of a sampling tube into one end of a piece of flexible hose.
- ☐ Connect the other end of the flexible hose to the air sampling pump.
- ☐ Connect the end of a second flexible hose to a manual or electronic frictionless bubble-tube calibrator and the other end of this flexible hose to the inlet end of the sampling tube. This constitutes the calibration train.
- ☐ Aspirate the bubble solution into the calibration device to form a soap film (bubble).

Calibration Data Form:
Low-Flow, High-Flow, and Multi-Flow Air Sampling Pumps

Name and Location of Calibration: ______________________________

Calibration Conducted By: ______________________________ **Date Calibration Conducted:** ________

Calibration Instruments (Type/Manufacturer/Model): ______________________________

Air Sampling Pump (Type/Manufacturer/Model): ______________________________

Collection Medium In-Line: ______________ **Calibrator Volume:** ________cc

Air Temperature: ____°C **Air Pressure:** ____mm Hg **Relative Humidity:** ____%

Pump No.	Pre-Sampling Calibration Flow Rate (Q_{pre})					Post-Sampling Calibration Flow Rate (Q_{post})					Average Flow Rate
	Time (sec)				Q_{pre}	Time (sec)				Q_{post}	Q_{avg}
	T_1	T_2	T_3	T_{avg}	L/min	T_1	T_2	T_3	T_{avg}	L/min	L/min

Calibration Notes:

Figure 9.7 Calibration data form for low-flow air sampling pump.

Field Monitoring Data Form:
Integrated Monitoring for Gases and Vapors

Facility Name and Location: ______________________ **Contaminant Sampled:** ____________

Monitoring Conducted By: ______________________

Collection Medium: ____________ **Date Monitoring Conducted:** ____________

Method(s)(Source/Number/Name): ______________________

Monitoring Instruments (Type/Manufacturer/Model): ______________________

Air Temperature: ____°C **Air Pressure:** ____mm Hg **Relative Humidity:** ____%

Identification of Personnel or Area	Field Sample	Pump No.	Pre-Sample Calibration Flow Rate	Post-Sample Calibration Flow Rate	Average Flow Rate (Q)	Sample Start-Time	Sample Stop-Time	Total Sample Time (T)	Volume Sampled (Vol)

Field Notes:

Figure 9.8 Field sampling data form for organic vapors.

Laboratory Analysis Data Form:
Gas Chromatography Analysis of Organic Analytes

Monitored Facility Name and Location: ______________________ **Contaminants Sampled:** ______________

Monitoring Conducted By: ______________________ **Date Monitoring Conducted:** ______________

Laboratory Facility Name and Location: ______________________ **Contaminants Analyzed:** ______________

Analysis Conducted By: ______________________ **Date Analysis Conducted:** ______________

Method(s)(Source/Number/Name): ______________________

Gas Chromatograph (Type/Manufacturer/Model): ______________________

Carrier Gas: ________ **Column:** ________ **Detector:** ________

Fuel Gas: ________ **Detector Gas:** ________ **Injection Temperature:** ____ °C **Column/Oven Temperature:** ____ ° C

Air Temperature: ____°C **Air Pressure:** ____mm Hg **Relative Humidity:** ____%

Laboratory Sample Identification No.	Field Sample Identification No.	Volume of Desorption Solvent (ml)	Volume Sample Aliquot (ml)	Retention Time (min)	Peak Area	Measured Amount of Analyte (mg)	Volume of Air Sampled (m^3)	Concentration of Air Contaminant (mg/m3)	Concentration of Air Contaminant (ppm)

Laboratory Notes:

Figure 9.9 Laboratory analysis data form for GC analysis of organic analytes.

- ☐ Record the time it takes the soap film (bubble) to traverse a known volume (e.g., 50 to 100 ml).
- ☐ If a manual bubble-tube is used, calculate the flow rate. Otherwise read the flow rate directly from the electronic bubble-tube calibrator.
- ☐ If the flow rate is not 150 cm^3/min ±5%, adjust the flow control on the pump and repeat the calibration procedure until desired flow rate is achieved.
- ☐ Once the desired flow rate is achieved (in this example 150 cm^3/min ±5%), repeat the calibration check two- to three-times and calculate the average pre-sampling flow rate.
- ☐ Disconnect the pump from the remainder of the calibration apparatus and turn "OFF." The calibrated air sampling pump is ready for use.

2. Prepare Sampling Media

- ☐ Label a solid sorbent (charcoal) sampling tube with an identification number.

3. Air Sampling for Nonpolar Aromatic Hydrocarbon Vapor (e.g., xylene, toluene)

- ☐ Get approval to conduct sampling in an area where airborne levels of nonpolar aromatic hydrocarbon vapors are likely to be detectable and measurable, and perform sampling accordingly. Alternatively, conduct a simulated sampling exercise and obtain a sample spiked with toluene or xylene from your instructor for later analysis.
- ☐ Complete a field data form, remembering to record name, location, person or area sampled, date, air temperature, air pressure, relative humidity, analyte (e.g., toluene, xylene), air sampling pump manufacturer and model, pump identification, sampling medium, sample identification, flow rate, start and stop times for sampling, sampling duration, and air volume sampled.
- ☐ Turn "ON" pump and let operate for 5 min prior to attaching flexible hose and sample collection medium.
- ☐ Place and secure pump in area where sample will be collected.
- ☐ Connect flexible hose to pump.
- ☐ Attach the free end of the flexible hose to a sample tube holder.
- ☐ Carefully use a sampling tube end-breaker and snap both ends from the sampling tube.
- ☐ Insert the sampling tube into the tube holder so that the inlet port of the tube (front section) is open to the contaminated atmosphere.
- ☐ Record the sample identification number and start time, and allow sample to run for approximately 2 h depending on conditions.
- ☐ After an acceptable sampling period has elapsed, turn OFF pump, record stop time, and calculate sample time (T [min]).
- ☐ Remove sample tube from the sampling train, cap both ends, and transport sample to the laboratory.

4. Post-sampling Calibration of Air Sampling Pump

- ☐ Repeat steps summarized in Method #1, except do not adjust the flow rate.
- ☐ The post-sampling flow rate (Q_{post}) should be within ±5% of the pre-sampling flow rate (Q_{pre}).
- ☐ Determine the average sampling flow rate (Q_{avg}) by averaging the pre- and post-sampling flow rates and express average flow rate as cubic centimeters per minute.

5. Analysis of Sample

- ☐ Using a file, carefully score the sorbent tube above the front section and snap open the tube.
- ☐ Use the forceps and carefully remove the front plug from the tube.
- ☐ Use the probe and gently push the plug that sits below the back section so that the medium in the tube can be emptied. Push the medium gradually and empty the front section into a vial labeled A. Remove the center plug that separates the sections and empty the back section into a vial labeled B.

- ☐ Add 1-ml CS_2 to each vial, cap immediately, and gently agitate for 30 min.
- ☐ Adjust GC as per manufacturer's specifications. Set injection temperature to 225°C, detector temperature at 225°C, and column temperature at 100°C.
- ☐ Check calibration of the GC and establish a standard curve by injecting 5-µl aliquots of prepared or commercially known standards for the analytes (e.g., toluene and xylene) into the GC, measure peak areas, and plot the standard curve.
- ☐ Manually or automatically, separately inject 5-µl aliquots of the sample from vials A and B into the GC.
- ☐ Review the chromatogram and identify and measure the analytes by recording RTs and curve areas relative to a standard curve of known concentrations vs. peak areas.
- ☐ Calculate airborne concentration of organic vapors in units of parts per million.
- ☐ Complete a laboratory analysis data form, remembering to record name, location, date, air temperature, air pressure, analytes (nonpolar organic [e.g., toluene, xylene]), GC manufacturer and model, sample identification, air volume sampled, and concentration.

UNIT 9 EXAMPLES

Example 9.1

A low-flow air sampling pump was calibrated using a manual frictionless bubble-tube. What was the flow rate (Q) in cubic centimeters per minute (cm³/min) if it took an average (n = 3 trials) of 30 sec (T) for the bubble to traverse a volume (V) of 50 cm³?

Solution 9.1

$$Q\ (\text{cm}^3/\text{min}) = \frac{\text{V}\ (cm^3)}{T\ (\text{sec})} \times \frac{60\,\text{sec}}{\text{min}}$$

$$= \frac{50\ cm^3}{30\ \text{sec}} \times \frac{60\,\text{sec}}{\text{min}}$$

$$= 100\ \text{cm}^3/\text{min}$$

Example 9.2

A sample was collected for 4 h (T min) using a low-flow sampling pump at an average flow rate (Q) of 100 cm³/min. What was the volume (V[m³]) of air sampled by the pump (Note: 4 hours = 240 min)?

Solution 9.2

$$\text{V}\ (m^3) = Q\ (\text{cm}^3/\text{min}) \times T\ (\text{min}) \times \frac{1\ m^3}{10^6\,\text{cm}^3}$$

$$= 100\ \text{cm}^3/\text{min} \times 240\ \text{min} \times \frac{1\ m^3}{10^6\,\text{cm}^3}$$

$$= 0.024\ m^3$$

Example 9.3

A 0.024 m^3 volume of air was sampled for benzene vapor (MW = 78 g/mol) using a low-flow air sampling pump connected to a standard charcoal tube as an adsorbent medium. Laboratory analysis of the front and backup sections of the charcoal tube using GC indicated that a total of 550 µg C_6H_6 was detected and measured. What was the concentration (ppm) of benzene vapor in the air during the sampling period, assuming normal temperature (25°C) and pressure (760 mmHg) (Note: 550 µg = 0.55 mg)?

Solution 9.3

$$C_6H_6\ (ppm) = \frac{C_6H_6\ (mg)}{Air\ Volume\ (\mathrm{m}^3)} \times \frac{24.45\ \mathrm{l\ Vapor / mol}\ C_6H_6}{78\ \mathrm{g / mol}\ C_6H_6} \times \frac{1\ m^3}{10^3\ l} \times \frac{1\ g}{10^3\ mg} \times 10^6$$

$$= \frac{0.55\ \mathrm{mg}}{0.024\ \mathrm{m}^3} \times \frac{24.45\ \mathrm{l / mol}}{78\ \mathrm{g / mol}} \times \frac{1\ m^3}{10^3\ l} \times \frac{1\ g}{10^3\ mg} \times 10^6$$

$$= 7.2\ \mathrm{ppm}$$

UNIT 9 CASE STUDY: SAMPLING AND ANALYSIS OF AIRBORNE ORGANIC VAPORS

OBJECTIVES

Upon completion of Unit 9, including sufficient reading and studying of this and related reference material, and review of the short case study and hypothetical industrial hygiene sampling and analytical data presented below, learners will be able to:

- Calculate the concentrations and time-weighted averages (TWAs).
- Interpret the data and results.
- Concisely summarize the information in a report.
- Retain the general scientific and technical principles and concepts related to the topic.

INSTRUCTIONS

After reading the short case study, locate and read the applicable NIOSH method for sampling and analysis (http://www.cdc.gov/niosh/nmam/nmammenu.html). Then, review the data shown in the tables below and perform the necessary calculations to complete the tables (assume normal temperature pressure). Review and interpret the results and compare calculated TWAs to applicable occupational exposure limits (e.g., Occupational Safety and Health Administration permissible exposure limits [OSHA-PELs], American Conference of Governmental Industrial Hygienists threshold limit values [ACGIH-TLVs],, and NIOSH recommended exposure limits [NIOSH-RELs]). Also, prepare responses to the questions that follow the data section. Finally, prepare a fictional, but representative, report following the outline or similar format shown in Appendix B.

CASE STUDY

A small factory produces solvent-based paints. The major operation involves one worker per shift manually emptying drums of volatile organic solvents, such as toluene, and bags of various powdered additives, including color pigments, into large mixers. Due to the relatively high vapor pressures and volatility of the organic solvents, there is concern that potential personal exposure to airborne vapor is unacceptable. There is no local exhaust ventilation adjacent to the mixers nor any personal protective equipment worn. Accordingly, personal integrated air sampling was conducted by a certified industrial hygienist (CIH). Following sampling, the prepared labeled samples and field blanks were capped prior to sending them to an American Industrial Hygiene Association (AIHA) accredited laboratory for analysis. Sampling and analysis for airborne toluene vapor followed NIOSH Method #1500. Calibration, field, and laboratory data are summarized in the tables below.

DATA AND CALCULATIONS

Table 9.1 Calibration and Field Data

Name of Personnel or Area	Sample No.	Analytes	Avg. Pre-sample Q (cm^3/min)	Avg. Post-sample Q (cm^3/min)	Avg. Q (cm^3/min)	Sample Start Time	Sample Stop Time	Sample Duration (min)	Air Volume Sampled (l)
	132		Pump No. 342	Pump No. 342		1500	1700		
	126					1700	1900		
Tim P.	136	Toluene				1900	2100		
	128		50	52		2100	2300		
Blank	105	Toluene	NA	NA	N/A	N/A	N/A	N/A	NA

N/A = Not applicable.

Table 9.2 Lab Data

Name of Personnel or Area	Sample No.	Analytes	Mass Analyte Detected Front (mg)	Mass Analyte Detected Back (mg)	Mass Analyte Detected Total (mg)	Air Volume Sampled (m^3)	Conc. Analyte (mg/m^3)	Conc. Analyte (ppm)
	132		1.9	0.1				
	126		2.5	<LOD				
Tim P.	136	Toluene	2.9	<LOD				
	128		3.3	<LOD				
Blank	105	Toluene	<LOD	<LOD	<LOD	N/A		

LOD = Limit of detection.

N/A = Not applicable.

Table 9.3 Calculated Time-Weighted Averages

Name of Personnel or Area	Sample No.	Analytes	8-h TWA (ppm)
	132	Toluene	
T. Pender	126		
	136		
	128		
Blank	105	Toluene	

QUESTIONS

- Were the average pre- and post-sampling flow rates within ±5% of each other?
- Was the total sample volume collected within the min-max range specified in the applicable NIOSH method?
- What is the minimum number of field blanks that should be submitted to the laboratory, based on the number of samples collected and the applicable NIOSH method?
- What considerations or assumptions did you make when calculating the TWAs? For example, did you assume the workers took breaks (e.g., lunch) in areas outside the immediate area? Would you continue monitoring while workers are on breaks, even if outside the potential exposure area? If not, what if the actual sample collection period was less than 8 h (480 min)?
- What is the estimated limit of detection (LOD) for the analyte as per the applicable NIOSH method?
- How did the calculated TWAs compare to applicable occupational exposure limits, such as the OSHA-PEL, ACGIH-TLV, and NIOSH-REL?
- What was the basis for the established occupational exposure limits? (Learners and practitioners alike are strongly encouraged to obtain and read applicable information found in resources such as the NIOSH *Criteria Documents* and ACGIH *Documentation of the TLVs and BEIs*.)
- Are there other NIOSH or OSHA air sampling and analytical methods for the same agent? If so, what are the reference numbers and what are the differences relative to sample collection (e.g., medium used) and analysis (e.g., analytical instrument used)?

REPORT

Prepare a concise report (see Appendix B for outline of sample format) that includes typed text and tables (and figures if you desire). Your report needs to include a section on calibration (how you calibrate pump with representative media), sample collection (instruments, media, and NIOSH method used), and analysis (instrument and NIOSH method used). In other words, write your short field report so the reader knows what was done to prepare for sampling and analysis for airborne organic vapors. Also, summarize, interpret, and discuss the results; state your conclusions; and make applicable recommendations. Remember to add a bibliography at the end of the report citing any applicable book, journal, and Internet references that you consult. Finally, append a copy of the applicable NIOSH methods.

UNIT 10

Evaluation of Airborne Inorganic and Organic Gases, Vapors, and Mists: Integrated Personal and Area Monitoring Using an Air Sampling Pump with a Liquid Absorbent Medium

LEARNING OBJECTIVES

At the completion of Unit 10, including sufficient reading and studying of this and related reference material, learners will be able to correctly:

- Name, identify, and assemble the components of a common sampling train, including the specific sampling medium, used for conducting integrated sampling of airborne inorganic and organic gases and vapors using liquid absorbent media.
- Name, identify, and assemble the applicable calibration train that uses a primary standard for measuring flow rate of a multi- or high-flow air sampling pump.
- Calibrate a multi- or high-flow air sampling pump, with the applicable representative sampling medium in-line, using a manual or electronic frictionless bubble-tube.
- Summarize the principles of sample collection of airborne inorganic and organic gases and vapors using liquid absorbent media.
- Conduct integrated sampling of airborne inorganic and organic gases and vapors using liquid absorbent media.
- Prepare samples for analysis of collected gas or vapor using ultraviolet or visible (UV/Vis) light spectrophotometry.
- Name and identify the components of a UV/Vis spectrophotometer.
- Name, identify, and assemble the components necessary for analysis of samples for inorganic and organic gases and vapors using a UV/Vis spectrophotometer.
- Calibrate a UV/Vis spectrophotometer and establish a standard curve using known concentrations of an organic analyte.
- Summarize the principles of sample analysis of liquid absorbents used for collection of airborne inorganic and organic gases and vapors.
- Conduct analysis of a sample for gas or vapor using a UV/Vis spectrophotometry.
- Perform applicable calculations and conversions related to pre- and post--sampling flow rate of an air sampling pump, average flow rate of an air sampling pump, sampling time, sampled air volume, weight of inorganic or organic analyte collected, and concentration of inorganic and organic gas and vapor in units of parts per million (ppm).
- Record all applicable calibration, sampling, and analytical data using calibration, field sampling, and laboratory analysis data forms.

OVERVIEW

Although solid adsorbent media are preferred for conducting integrated sampling of gases and vapors, some compounds are not collected efficiently using these media. The solid absorbents, such as activated charcoal and silica gel, are used most frequently to collect relatively nonreactive gases or vapors. Liquid absorbents are used for relatively more soluble and reactive gases and vapors and for compounds that are soluble in the medium. As a result, liquid absorbents are used as an alternative collection media for organic gases and vapors, as well as some inorganic gases and mists. The use of industrial hygiene air sampling methods requiring use liquid absorbents, however, continues to steadily decline.

Unlike adsorption where gas and vapor molecules bond to the surface of a solid medium, absorption involves dissolution of molecules of gases and vapors into a liquid medium (Figure 10.1). The liquid absorbents are contained in specialized sampling devices during sampling. The purpose of the sampling devices is to enhance aeration of the liquid absorbent with the sampled air so that increased surface area and mixing are available. In turn, increased contact between the gas and vapor molecules and the liquid absorbent enhances the efficiency of collection of the contaminant.

COMMON LIQUID ABSORPTION DEVICES

Midget impingers and fritted bubblers are the most common liquid absorption devices used for air sampling in the occupational environment (Figure 10.2). Their use is generally limited to the collection of more soluble or reactive gases, vapors, and mists, both inorganic and organic.

Midget impingers are columnar glass devices that provide a reservoir for liquid absorption reagents. Inserted in the reservoir is a hollow glass inlet tube, opened at each end and tapered at the bottom. The tube is attached and extends through the top of the reservoir. When the reservoir is filled with 10 to 20 ml of a liquid medium, the tapered end of the tube is immersed in the fluid. The tube serves as a conduit for the gaseous air contaminant to enter the impinger and mix with and dissolve into the liquid absorbent.

Midget bubblers are designed very much like impingers, except that the glass inlet tube is not tapered. The end of the hollow tube in a midget bubbler is a porous piece of glass with a sponge-like appearance. This structure is called a frit and is immersed in the liquid absorbent. When air is drawn through the fritted glass at the end of the tube, smaller and more bubbles are generated relative to the midget impinger. The actual size of the bubbles depends on the nature of the liquid and the diameter of the frit pores. Frits range from fine to extra coarse, depending on the number of pores per unit area. The more coarse the frit, the more rapid the flow. The small bubbles generated by a frit allow increased surface area for contact and mixing of gases and vapors with the absorbent. These characteristics can result in increased collection efficiency. Midget bubblers are generally used to collect less reactive gases or vapors that might not be collected as efficiently using midget impingers.

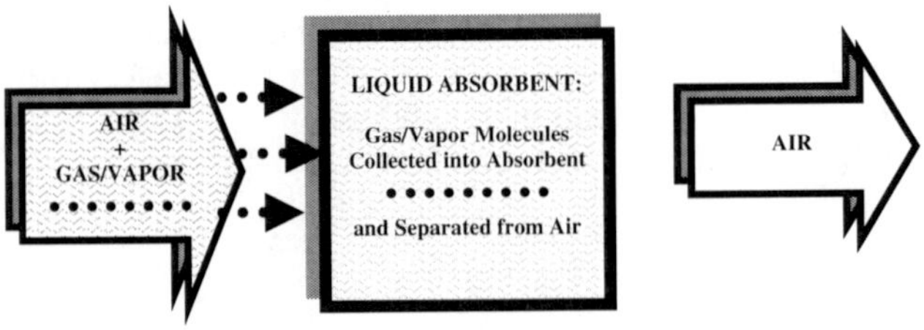

Figure 10.1 Separation of organic vapor samples from air via absorption into a liquid medium.

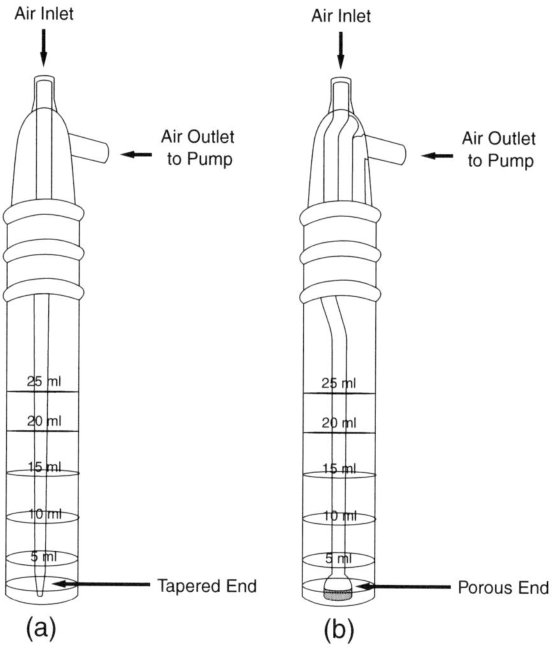

Figure 10.2 Glass midget (a) impinger and (b) bubbler.

Treated and other specialty adsorbents (see Unit 9) have been developed to replace the more cumbersome methods requiring liquid absorbents. Also, as discussed in Unit 9, alternative sampling devices called passive monitors also have been developed; they are continually being perfected for sampling inorganic and organic gases and vapors that have been traditionally sampled using impingers and bubblers containing liquid absorbent.

PRECAUTIONS

A major disadvantage of using midget impingers and bubblers is that the liquid absorbent can spill or be pulled into the air sampling pump. As a result, it is important that the impingers be positioned vertically during sampling. Holster-style devices have been designed to assist in maintaining the impinger in a vertical position. In addition, so-called spill-proof devices are also available. A spill trap also can be used, such as an empty impinger in-line between the pump and the impinger that contains the liquid medium. It is also possible for volatile liquid absorbents to generate vapors which can be pulled into the pump. When these types of reagents are used, a vapor trap, such as a charcoal tube for nonpolar or a silica gel tube for polar solvents, in-line between pump and impinger may be warranted. There are several commercially available liquid and vapor traps.

Adequate contact time between the air bubbles and the absorbent is important to increase collection efficiency. This is controlled by operating the air sampling pumps at a flow rate less than 1 l/min. Collection efficiency can also be enhanced by increasing the volume of the liquid through which the bubbles have to pass. This can be accomplished by adjusting the volume in an individual impinger or using two reagent-filled impingers in-line. Finally, the liquid absorbent must be an acceptable solvent for the gas and vapor that is to be collected. In general, the solubility of the absorbent will be compatible with the solubility of the gas or vapor, without concern for interference with the analytical method.

SAMPLING

The air is actively drawn into the impinger or bubbler by means of an air sampling pump that is connected via a flexible hose to a glass outlet tube that projects horizontally from the top of the device (Figure 10.3). As the air is pumped from the atmosphere and into and through the vertical glass inlet tube, bubbles are formed as the air enters the liquid medium. The combination of the bubbles and mixing facilitates collection or trapping of the gas or vapor molecules in the liquid absorbent. A brief outlined summary of sample collection follows.

(i) Calibration

- An air sampling pump is calibrated using a manual or electronic calibrator with midget impinger or bubbler plus optional vapor or liquid traps in-line (Figure 10.4).
- Pre- (Q_{pre}) and post-sampling (Q_{post}) flow rates are determined by measuring the average time (T_{avg} sec) based on three trials ($\Sigma T_{1,2,3}/3$) for a bubble to traverse a specific volume (V [cm^3]) of the bubble-tube.

$$Q\,(1/\min) = \frac{V\ (cm^3)}{T_{avg}(sec)} \times \frac{60\ sec}{1\,min} \times \frac{1\,l}{10^3 cm^3} \qquad (10.1)$$

- The average flow rate (Q_{avg}) is based on the average of pre- (Q_{pre}) and post-sampling (Q_{post}) flow rates.

$$Q_{avg} = \frac{Q_{pre} + Q_{post}}{2} \qquad (10.2)$$

- The typical flow rate is 1 l/min.

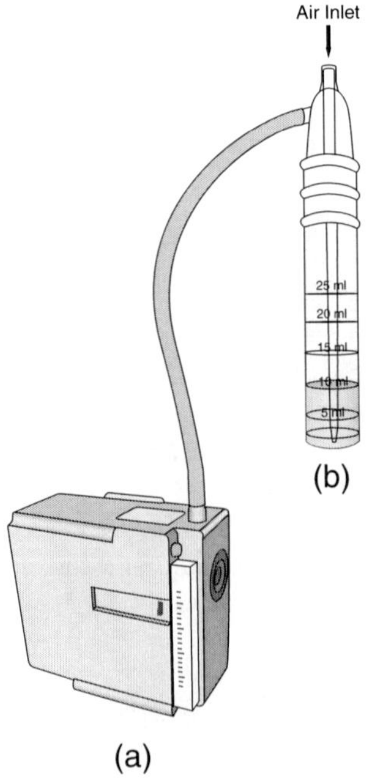

Figure 10.3 Sampling train for organic vapors composed of (a) a high-flow pump connected with flexible hose to (b) a glass midget impinger filled with liquid absorbent medium.

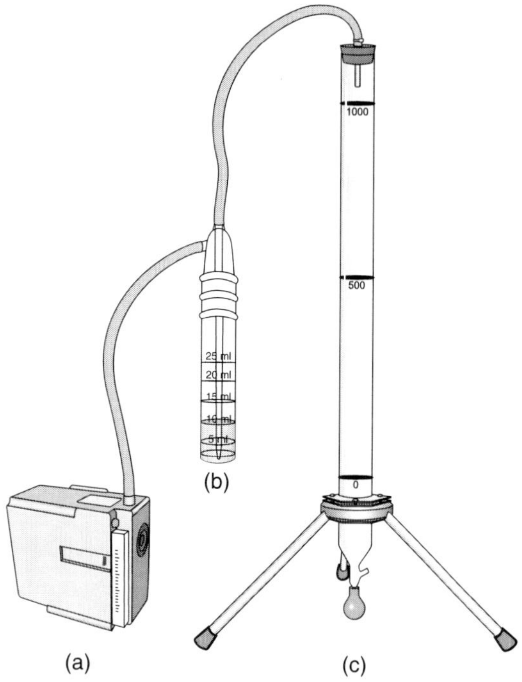

Figure 10.4 Calibration train for organic vapors composed of (a) a high-flow pump connected with flexible hose to (b) a glass midget impinger filled with liquid absorbent medium in-line with (c) a frictionless bubble-tube (vapor or liquid trap not shown).

(ii) Preparation for Sampling

- A midget impinger or bubbler is filled with a specific volume (10 to 20 ml) of known liquid absorbent suitable for collecting or trapping the air contaminant.
- A sampling train is assembled and consists of multi- or high-flow pump, 2-ft 1/4-in. i.d. flexible hose, liquid and/or vapor trap, 0.5-ft. 1/4-in. i.d. flexible hose, and midget impinger or bubbler in a holster or other suitable support.

(iii) Conducting Sampling

- A midget impinger or bubbler is attached within the breathing zone, and the air sampling pump is clipped to the belt of the worker for personal sampling. The sampling train is positioned and secured in a specific location for area sampling.
- The pump is turned "ON" and start time is recorded.
- After a specified time, the pump is turned "OFF" and stop time is recorded.
- The liquid absorbent (and dissolved contaminant) contained within the midget impinger or bubbler is decanted into a labeled vial, capped, and transported to laboratory for analysis.

ANALYSIS

UV/Vis spectrophotometry is the method of analysis most often utilized for contaminants collected in liquid absorbents (Figure 10.5). The liquid absorbent plus the dissolved contaminant that was removed from the impinger or bubbler are mixed with reagents at a laboratory. The reaction of the analytical reagents and the contaminant yield a product or analyte that will maximally absorb a wavelength of light in either UV or Vis spectrum. The wavelength of light that is absorbed

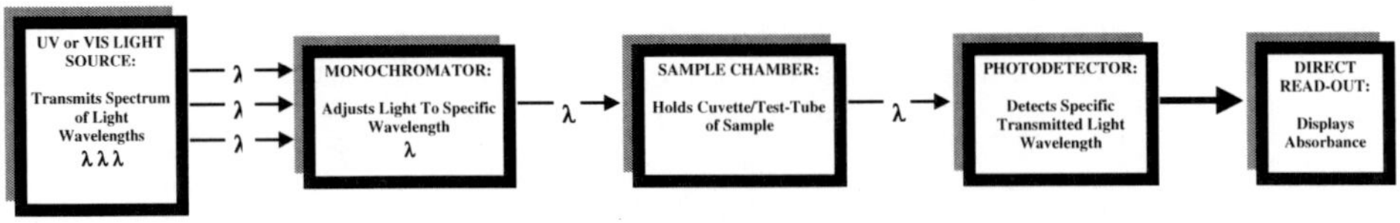

Figure 10.5 Components of a UV/Vis spectrophotometer.

maximally is selected during UV or Vis spectrophotometry. A detailed description of UV/Vis spectrometry is beyond the scope of this book. A brief outlined summary of the components and principles follows.

(i) Components

- Light bulb: provides source of light such as a tungsten or tungsten-iodine bulb for Vis spectra (and infrared [IR]) and deuterium bulb for ultraviolet (UV) spectra.
- Monochromator: adjusts the wavelength of light by screening out the spectrum of wavelengths, except the wavelength that is absorbable by the analyte in the sample. Wavelengths can be within the UV (~180 to 380 nm) or the Vis (~380 to 750 nm) spectrum. Some spectrophotometers provide an IR (750 to 15,000 nm) spectrum.
- Sample chamber: the compartment where the sample container (cuvette or test tube) is placed for illumination with the specific wavelength of light.
- Detector: the photocell intercepts light energy and converts it into electrical energy. The intensity of light detected is related to the electrical energy generated.
- Readout: an electrical signal is relayed from the detector to the readout device which indicates transmittance (T) and absorbance (Abs). Commonly interfaced with a computer.

(ii) Principle of Operation

- A specific spectrum (UV or Vis) and wavelength of light are selected relative to the contaminant or analyte in the sample.
- A sample solution is placed in a special cuvette into the sample chamber.
- A specific wavelength of light is directed to the sample chamber where analyte molecules absorb the light energy. A sample blank (distilled water or reagent solution without contaminant) will permit 100% transmittance of light, but zero absorbance. If present, however, the analyte will absorb the wavelength of light in direct relation (linearly) to its concentration (Beer's law).
- Light not absorbed by the sample analyte exits the sample chamber to the detector, generating an electrical signal to the readout.
- The intensity of the light detected is converted to display as Abs on a direct readout.

(iii) Determination of Concentration of Inorganic or Organic Analyte

- Beer's law suggests proportionality between Abs and concentration of analyte (c).
- A gradient of concentrations of known standard analyte is prepared and analyzed via UV/Vis spectrophotometry to generate a standard curve of absorbance vs. concentration by plotting absorbance vs. concentration of analyte ($Organic_{std}$) in units of micrograms per milliliter (μg/ml) (Figure 10.6).
- Absorbance reading for the analyzed sample (unknown concentration) is compared to the readings plotted for the analyzed standards (known concentrations). The concentration of analyte (μg/ml) in the sample solution that was aspirated into and detected and measured by the UV/Vis spectrophotometer is extrapolated from the standard curve.
- The amount (weight) of the analyte (e.g., HCHO) in the analyzed sample is calculated by multiplying the concentration of HCHO in the sample solution (C_{samp} [μg/ml]), based on the

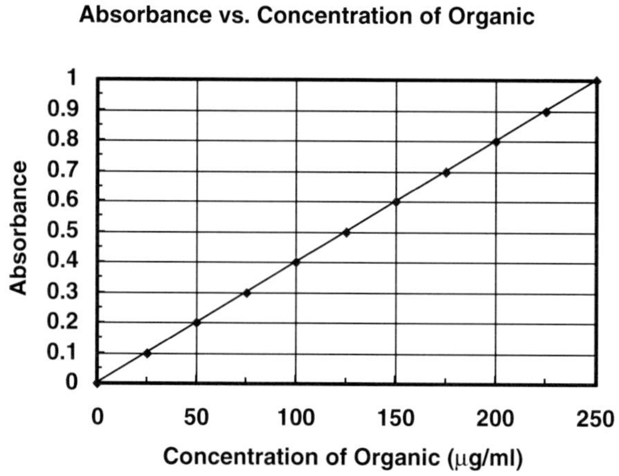

Figure 10.6 Standard curve for UV/Vis spectrophotometric analysis of known concentrations of organic analyte in solution.

extrapolation from the standard curve, by the final volume of sample solution (V_{samp}[ml]) used to prepare the liquid sampling absorbent and sample prior to analysis.

$$\text{Wt. Organic (mg)} = V_{samp}\ (\text{ml}) \times C_{samp}\ [(\text{Organic}\ [\mu\text{g}] / \text{Solution}\ [\text{ml}])] \times \frac{1\ \text{mg}}{10^3\ \mu\text{g}} \quad (10.3)$$

- The sampled volume of air is determined by multiplying sampling time (T [min]) by average sampling flow rate (Q [l/min]). Convert liters (l) to cubic meters (m^3).

$$\text{Air Volume}\ (m^3) = T\ (\text{min}) \times Q_{avg}\ (\text{l}/\text{min}) \times \frac{1\ m^3}{10^3\ \text{l}} \quad (10.4)$$

- The concentration of analyte is determined by dividing the weight of analyte (mg) by the sampled volume of air (m^3). The concentration is expressed in units of milligrams of analyte (e.g., HCHO) per cubic meter of air (mg/m^3).

$$Organic\ (\text{mg}/\text{m}^3) = \frac{Organic\ (mg)}{Air\ \text{Volume}\ (\text{m}^3)} \quad (10.5)$$

- Convert milligrams per cubic meter (mg/m^3) to parts per million (ppm).

$$Organic\ (\text{ppm}) = \frac{Organic\ (mg)}{Air\ \text{Volume}\ (\text{m}^3)} \times \frac{24.45\ \text{l}/\text{g-mol}}{\text{MW}\ (\text{g}/\text{g-mol})} \times \frac{1\ m^3}{10^3\ l} \times \frac{1\ g}{10^3\ mg} \times 10^6 \quad (10.6)$$

- Note: The adjusted concentration (C) can be calculated by subtracting the weight of organic measured in a reagent blank (B) from the sample reagent (S), dividing by air volume sampled (V), and also correcting for sampling efficiency (E) which is equal to ≤1 depending if 100% or less efficiency is attained.

$$C = \frac{S - B}{E\ V} \quad (10.7)$$

UNIT 10 EXERCISE

OVERVIEW

The exercise will provide the fundamental concepts for conducting integrated sampling for an organic vapor using a liquid absorbent medium. In addition, a common and related analytical method using Vis spectrophotometry is introduced. The exercise will focus on sampling and analysis of formaldehyde vapors. The material and methods are based on a modification of the National Institute for Occupational Safety and Health (NIOSH) Method #3500. Place a check mark (✔) in the open box (☐) when you have obtained applicable material and completed the steps for sampling and analytical methods.

MATERIAL

1. Calibration

- ☐ Midget impinger filled with 20-ml distilled water
- ☐ Multi- or high-flow air sampling pump
- ☐ Impinger holder
- ☐ Two 2-ft lengths and one 6-in. length of ¼-in. i.d. flexible hose
- ☐ Manual or electronic frictionless bubble-tube calibrator
- ☐ Soap solution
- ☐ Stopwatch (if manual calibrator is used)
- ☐ Calibration data form (Figure 10.7)

2. Sample Collection

- ☐ Midget impinger filled with 20 ml 1% sodium bisulfite; 10.0 g $NaHSO_3$ is dissolved in deionized H_2O and diluted to 1 l
- ☐ Glass vials with caps for transfer of liquid absorbent
- ☐ Multi- or high-flow air sampling pump
- ☐ Impinger holder
- ☐ One 2-ft length and one 6-in. length of ¼-in. i.d. flexible hose
- ☐ Field sampling data form (Figure 10.8)

3. Sample Analysis

- ☐ Chromotropic acid: Dissolve 1-g 4,5-dihydroxy-2,7-naphthalene disulfonic acid disodium salt in deionized H_2O and dilute to 100 ml. Store in amber bottle and prepare weekly.
- ☐ Formaldehyde (HCHO) stock solution (1 mg/ml): Dissolve 4.4703 g sodium formaldehyde bisulfite ($HOCH_2SO_3Na$) in deionized H_2O and dilute to 1 l. Store in amber bottle in a cool place and prepare weekly.
- ☐ Known concentration solutions: Dilute aliquots of HCHO stock solution with 1% sodium bisulfite solution to establish a concentration gradient ranging from 0 to 20 μg/ml. Store in amber bottle in a cool place and prepare from fresh (≤1 week old) stock solution.
- ☐ Reagent blank: 4 ml 1% sodium bisulfite solution +0.10 ml chromotropic acid +6 ml sulfuric acid
- ☐ Water bath.
- ☐ UV/Vis spectrophotometer set at visible wavelength equal to 580 nm.
- ☐ Spectrophotometer cuvettes or test tubes.
- ☐ Pipettes.
- ☐ Laboratory analysis data form (Figure 10.9).

Calibration Data Form:
Low-Flow, High-Flow, and Multi-Flow Air Sampling Pumps

Name and Location of Calibration: ____________________

Calibration Conducted By: ____________________ Date Calibration Conducted: ________

Calibration Instruments (Type/Manufacturer/Model): ____________________

Air Sampling Pump (Type/Manufacturer/Model): ____________________

Collection Medium In-Line: ____________ Calibrator Volume: ________cc

Air Temperature: ____°C Air Pressure: ____mm Hg Relative Humidity: ____%

Pump No.	Pre-Sampling Calibration Flow Rate (Q_{pre})					Post-Sampling Calibration Flow Rate (Q_{post})					Average Flow Rate
	Time (sec)				Q_{pre}	Time (sec)				Q_{post}	Q_{avg}
	T_1	T_2	T_3	T_{avg}	L/min	T_1	T_2	T_3	T_{avg}	L/min	L/min

Calibration Notes:

Figure 10.7 Calibration data form for high-flow air sampling pump.

Field Monitoring Data Form:
Integrated Monitoring for Gases and Vapors

Facility Name and Location: ______________________ **Contaminant Sampled:** ____________

Monitoring Conducted By: ______________________

Collection Medium: ____________ **Date Monitoring Conducted:** ____________

Method(s)(Source/Number/Name): ______________________

Monitoring Instruments (Type/Manufacturer/Model): ______________________

Air Temperature: ____°C **Air Pressure:** ____mm Hg **Relative Humidity:** ____%

Identification of Personnel or Area	Field Sample	Pump No.	Pre-Sample Calibration Flow Rate	Post-Sample Calibration Flow Rate	Average Flow Rate (Q)	Sample Start-Time	Sample Stop-Time	Total Sample Time (T)	Volume Sampled (Vol)

Field Notes:

____

____

Figure 10.8 Field sampling data form for organic vapors.

Laboratory Analysis Data Form:
UV/Vis Spectrophotometry Analysis of Inorganic and Organic Analytes

Monitored Facility Name and Location: ______________________ **Contaminants Sampled:** ____________

Monitoring Conducted By: ______________________ **Date Monitoring Conducted:** ____________

Laboratory Facility Name and Location: ______________________ **Contaminants Analyzed:** ____________

Analysis Conducted By: ______________________ **Date Analysis Conducted:** ____________

Method(s)(Source/Number/Name): ______________________

UV/Vis Spectrophotometer (Type/Manufacturer/Model): ______________________

Light Source: ________ **Wavelength:** ________ **Path Length:** ________ **Analyte Extinction Coefficient:** ________

Analytical Method: ______________________

Air Temperature: ____°C **Air Pressure:** ____mm Hg **Relative Humidity:** ____%

Laboratory Sample Identification No.	Field Sample Identification No.	Volume Prepared Sample (ml)	Volume Sample Aliquot (ml)	Absorbance	Measured Amount of Analyte (mg)	Volume Air Sampled (m^3)	Concentration of Air Contaminant (mg/m3)	Concentration of Air Contaminant (ppm)

Laboratory Notes:

Figure 10.9 Laboratory analysis data form for UV/Vis spectrophotometry analysis of organic analytes.

METHOD

1. Pre-sampling Calibration of Sampling Pump

- ☐ Remove pump from charger, turn "ON," and allow it to run for 5 min.
- ☐ Complete a calibration form, remembering to record name, location, date, temperature, pressure, pump manufacturer and model, calibration apparatus, times for three trials, and average flow rate for pre- and post-sampling calibrations.
- ☐ Obtain a model sampling medium for use during calibration of the air sampling pump. This should consist of a midget impinger filled with 20-ml H_2O.
- ☐ Connect one end of the flexible hose to the glass tube arm that projects from the side of the impinger.
- ☐ Connect the other end of the flexible hose to the air sampling pump.
- ☐ Connect one end of the second flexible hose to the inlet port (vertical hollow glass tube) of the impinger.
- ☐ Connect the remaining end of the second flexible hose to a manual or electronic frictionless bubble-tube calibrator. This constitutes the calibration train.
- ☐ Aspirate the bubble solution into the calibration device to form a soap film (bubble).
- ☐ Record the time it takes the soap film (bubble) to traverse a known volume (e.g., 500 to 1000 ml).
- ☐ If a manual bubble-tube is used, calculate the flow rate. Otherwise read the flow rate directly from the electronic bubble-tube calibrator.
- ☐ If the flow rate is not 0.5 l/min ±5%, adjust the flow control on the pump and repeat calibration procedure until desired flow rate is achieved.
- ☐ Once the desired flow rate is achieved (in this example 0.5 l/min ±5%), repeat the calibration check two to three times and calculate the "average pre-sampling flow rate."
- ☐ Disconnect the pump from the remainder of the calibration apparatus and turn "OFF." The calibrated air sampling pump is ready for use.

2. Prepare Sampling Media

- ☐ Label an impinger with a sample identification number.
- ☐ Add 20-ml 1% sodium bisulfite solution as the sampling medium.
- ☐ Connect the impinger that contains the reagent to an empty impinger used for a liquid trap.

3. Air Sampling for Formaldehyde Vapors

- ☐ Get approval to conduct sampling in an area where airborne levels of formaldehyde vapors are likely to be detectable and measurable, and perform sampling accordingly. Alternatively, conduct a simulated sampling exercise and obtain a sample spiked with formaldehyde from your instructor for later analysis.
- ☐ Complete a field sampling data form, remembering to record name, location, person or area sampled, date, temperature, pressure, analyte (e.g., formaldehyde vapor), air sampling pump manufacturer and model, pump identification, sampling medium, sample identification, flow rate, start and stop times for sampling, sampling duration, and air volume sampled.
- ☐ Turn "ON" pump and let run for 5 min.
- ☐ Place and secure pump in area where sample will be collected.
- ☐ Connect flexible hose to pump.
- ☐ Attach the free end of the flexible hose to the glass tube arm that projects from the side of an empty impinger.
- ☐ Connect a short length of flexible hose to the inlet glass tube of the empty impinger. Connect the other end of the flexible hose to the glass tube arm that projects from the impinger containing reagent.
- ☐ Insert the midget impinger that contains the reagent into a holder so that the inlet port of the impinger is open to the contaminated atmosphere. The impinger must be kept vertical.

- ☐ Record the sample number and start time, and allow sample to run for approximately 2 h depending on conditions.
- ☐ After an acceptable sampling period has elapsed, turn "OFF" pump and record stop time.
- ☐ Remove midget impinger from the sampling train. Transfer the contents into a labeled glass vial, cap tightly, and transport sample to laboratory. During transfer, rinse the absorbing liquid adhering to the outside and inside of the stem directly into the impinger or bubbler vial with a small volume (1 to 2 ml) of the sampling reagent.

4. Post-sampling Calibration of Air Sampling Pump

- ☐ Repeat steps summarized in Method #1, except do not adjust the flow rate of the pump.
- ☐ The post-sampling flow rate (Q_{post}) should be within ±5% of the pre-sampling flow rate (Q_{pre}).
- ☐ Determine average sampling flow rate (Q_{avg}) by averaging the pre- and post-sampling flow rates and express average flow rate as liters per minute.

5. Analysis of Sample

- ☐ Adjust the spectrophotometer to 0% absorbance (Abs) using a deionized H_2O blank.
- ☐ Read Abs_{580} of reagent blank.
- ☐ Analyze gradient of prepared standard known solutions to establish a calibration or standard curve (concentration vs. absorbance).
- ☐ Under a laboratory exhaust hood, mix 4 ml sample from +0.1 ml chromotropic acid +6 ml sulfuric acid (Caution: exothermic reaction) in a 25-ml flask and stopper.
- ☐ Heat the solution to 95°C for 15 minutes in a water bath and cool to room temperature.
- ☐ Read Abs_{580}.
- ☐ Run samples, measure absorbance, and compare against a standard curve of known concentrations vs. absorbance.
- ☐ Calculate the airborne concentration of formaldehyde in units of parts per million.
- ☐ Complete a laboratory analysis data form, remembering to record name, location, date, air temperature, air pressure, analytes (e.g., HCHO), UV/Vis spectrophotometer manufacturer and model, sample identification, air volume sampled, and concentration.

UNIT 10 EXAMPLES

Example 10.1

A high-flow air sampling pump was calibrated using a manual frictionless bubble-tube. What was the flow rate (Q) in liters per minute (l/min) if it took 30 sec (T) for the bubble to traverse a volume (V) of 500 cm^3?

Solution 10.1

$$Q\ (1/\min) = \frac{\mathrm{V}\ (cm^3)}{T\ (\sec)} \times \frac{60\ \sec}{\min} \times \frac{1\ l}{10^3\ cm^3}$$

$$= \frac{500\ cm^3}{30\ \sec} \times \frac{60\ \sec}{\min} \times \frac{1\ 1}{10^3\ cm^3}$$

$$= 1\ 1/\min$$

Example 10.2

A sample was collected for 8 h (T min) using a high-flow sampling pump at a flow rate (Q) of 1 l/min. What was the volume (V [m^3]) of air sampled by the pump (Note: 8 h = 480 min)?

Solution 10.2

$$\mathrm{V}\ (m^3) = Q\ (\mathrm{l/min}) \times T\ (\mathrm{min}) \times \frac{1\ m^3}{10^3 l}$$

$$= 1\ \mathrm{l/min} \times 480\ \mathrm{min} \times \frac{1\ m^3}{10^3 l}$$

$$= 0.480\ m^3$$

Example 10.3

A 0.480 m^3 volume of air was sampled for formaldehyde vapor using a high-flow air sampling pump connected to a glass midget impinger containing 20 ml 1% sodium bisulfite as a liquid absorbent medium. Laboratory analysis of the prepared sample using Vis spectrophotometry followed by extrapolation from a standard curve indicated that formaldehyde was present at 50 µg/ml. What was the concentration (ppm) of formaldehyde vapor in the air during the sampling period?

Solution 10.3

$$\mathrm{HCHO\ (ppm)} = \mathrm{V}\ (ml) \times \frac{\mathrm{HCHO/ml\ (mg)}}{Air\ Volume\ (m^3)} \times \frac{\mathrm{24.45\ Vapor/mol}}{\mathrm{30\ g/mol\ HCHO}} \times \frac{1\ m^3}{10^3\ 1} \times \frac{1\ g}{10^3\ g} \times 10^6$$

$$= 20\ ml \times \frac{0.05\ mg}{0.480\ m^3} \times \frac{24.45\ l/\mathrm{mol}}{30\ \mathrm{g/mol}} \times \frac{1\ m^3}{10^3\ 1} \times \frac{1\ g}{10^3\ \mathrm{mg}} \times 10^6$$

$$= 1.6\ ppm$$

UNIT 11

Evaluation of Airborne Combustible and Oxygen Gases: Instantaneous Area Monitoring Using a Combined Combustible and Oxygen Gas Meter

LEARNING OBJECTIVES

At the completion of Unit 11, including sufficient reading and studying of this and related reference material, learners will be able to correctly:

- Name and identify the common instrument used for conducting instantaneous monitoring for both airborne combustible gases and vapors and oxygen gas.
- Name, identify, and assemble the applicable calibration train that uses known concentrations of reference gases for calibrating a combined combustible and oxygen gas meter.
- Calibrate a combined combustible and oxygen gas meter using applicable known concentrations of reference gases.
- Summarize the principles of sample collection of airborne combustible gases and vapors and oxygen gas using a combined combustible and oxygen gas meter.
- Conduct instantaneous monitoring of airborne combustible gases and vapors and oxygen gas using a combined combustible and oxygen gas meter.
- Record all applicable calibration and sampling data using a field monitoring data form.

OVERVIEW

Two immediate threats to the health of workers are the presence of a combustible atmosphere and either an oxygen-deficient or oxygen-enriched atmosphere. Numerous organic gases and vapors act as combustible fuels if there is an adequate ratio of them to air. The combustible gas or vapor is a reducing agent and oxygen in air is an oxidizing agent. In the presence of an ignition source, an oxidation-reduction reaction involving fuel and air can be initiated. A subsequent chain reaction of oxidation-reduction reactions results in propagation of flame or an explosion. The chain reactions continue until the fuel or the oxygen is depleted or the combustion process is extinguished (e.g., cooled, smothered).

Most organic gases and vapors will combust or explode within a flammability range based on fuel to air ratios. The extremes of the flammability range are referred to as the "lower flammability or explosive limit" (LFL or LEL) and upper flammability or explosive limit (UFL or UEL). The LFL or LEL is the minimum ratio of fuel to air that will support propagation of flame; ratios less than the LFL are considered too lean. The UFL or UEL is the maximum ratio of fuel to air that will support propagation (spreading) of flame; ratios greater than the UFL are considered too rich.

For example, the solvent isobutyl acetate has an LFL of 1.3% and UFL of 10.5% and, accordingly, a flammability range between 1.3 to 10.5%.[1] The concentration of isobutyl acetate in air would be outside of its flammability range, therefore, when less than 1.3% or greater than 10.5%. When outside of the range, combustion with propagation of flame would not occur.

Oxygen gas is a normal and essential component in air at a concentration of approximately 20.8%. When levels decrease significantly below 20.8%, the atmosphere is designated oxygen deficient (e.g., <19.5% O_2). When levels increase significantly, the atmosphere is oxygen enriched (e.g., >23% O_2). Deficient and excessively enriched atmospheres can result in toxic effects to individuals present due to hypoxia and hyperoxia, respectively. Enriched atmospheres also increase the potential for initiation of combustion and explosion reactions.

SAMPLING AND ANALYSIS

Monitoring devices are available that instantaneously detect and measure the level of airborne combustible gas or vapor and oxygen gas. The most common devices for detecting and measuring combustible and oxygen gases are used for area monitoring. Although individual instruments are available to measure either combustible gases or vapors or oxygen gas separately, it is more common to use a combination unit that is designed to detect and measure both parameters. These combined devices are called combination combustible and oxygen gas meters or, more simply, combustible gas indicators or meters (Figures 11.1 and 11.2). Many of the combination meters also measure a toxic parameter such as carbon monoxide or hydrogen sulfide gases. These instruments are very important for hazardous materials, emergency response activities, and confined space entry monitoring and clearance activities.

Monitoring of combustible gas or vapor and oxygen gas is instantaneous and represents real-time measurements of the analytes. The devices are typically electronic active flow monitoring instruments that collect an air sample by automatically pumping air into the device. As the air flows through the instrument, combustible gas or vapor and oxygen gas interact with sensors or detectors for the respective analytes. The meters are generally calibrated for combustible gas or vapor using a known concentration of reference gas such as pentane or methane. Responses to other combustible gases or vapors actually detected and measured during monitoring must be adjusted via conversion

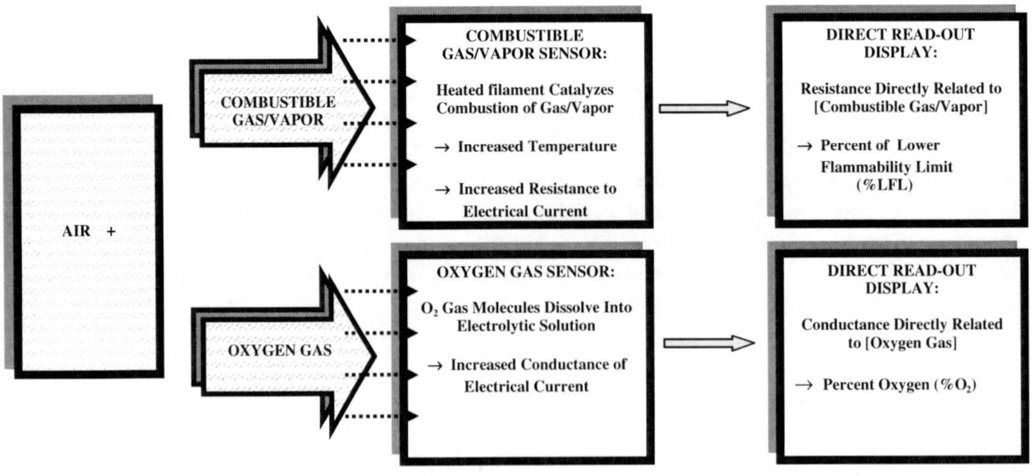

Figure 11.1 Schematic of combined combustible and oxygen gas meter.

[1] U.S. Department of Health and Human Services, Public Health Service, Centers for Disease Control, *NIOSH Pocket Guide to Chemical Hazards*, U.S. Government Printing Office, Washington, D.C., 2003.

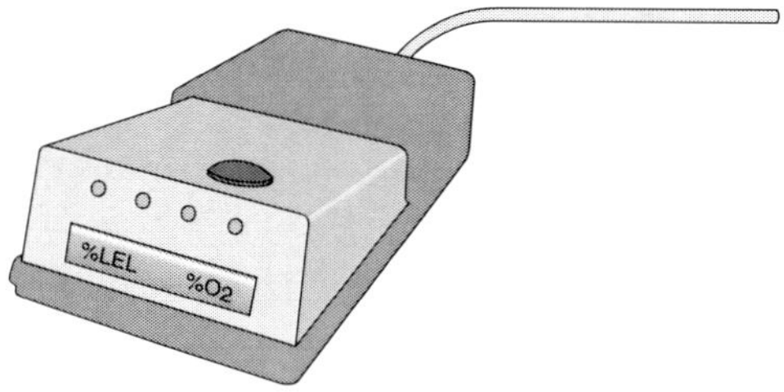

Figure 11.2 Combined combustible and oxygen gas meter.

factors to determine the exact concentration of a specific gas. For oxygen, some meters are calibrated using a known concentration (reference gas) of molecular oxygen gas. It is also common to calibrate an oxygen gas meter using outdoor (fresh) air. Reference gases used to calibrate the instruments are contained within compressed gas cylinders (Figure 11.3).

(i) Combustible Gas Meters

Combustible gas meters (CGMs) or indicators commonly contain a heated wire such as a platinum filament that internally ignites and catalyzes the combustible gas or vapor. The ignition of a combustible gas or vapor generates a detectable and measurable quantity of heat. This heat of combustion decreases the resistance of the filament in proportion to the concentration of the combustible gas or vapor present.

The readout of the meter indicates the concentration of combustible gases or vapors as percent of the LFL. For example, if the meter reading (display) is 8%, that means 8% of the LFL. If isobutyl acetate with an LFL of 1.3% was the only detectable vapor present during sampling, the actual concentration would be 0.08 multiplied by 1.3%, which equals 0.104% based on Equation 11.1.

$$\text{Concentration}_x = \text{Meter Display}_x \times \text{LFL}_x \qquad (11.1)$$

This actually represents a percentage of the LFL (or LEL) relative to the specific calibration gas. For example, in the case involving isobutyl acetate, a meter display of 8% (or as decimal fraction, 0.08) of the LFL may not have been the actual reading. Depending on the make and model of the CGM, a specific correction factor (CF) or response factor is assigned for specific flammable gases and vapors relative to the calibration gas. This is necessary since when a CGM is calibrated

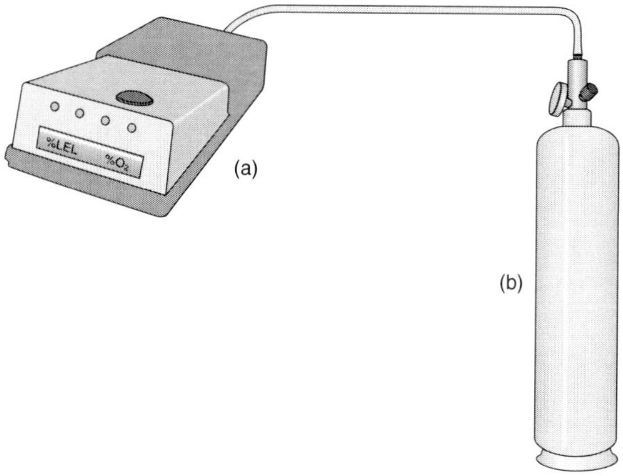

Figure 11.3 Calibration train for combined combustible and oxygen gas meter consisting of (a) a meter connected to (b) a cylinder of compressed reference gas.

using a specific calibration gas, for example methane gas, the instrument will respond to other flammable gases and vapors as though it is detecting and measuring the methane. Accordingly, the CGM readout display or measured value may be higher or lower than the actual value.

Continuing with the isobutyl acetate example, assume that the CGM was calibrated with pentane and that under these conditions the CF for isobutyl acetate is 1.5. Accordingly, the actual reading would be 1.5 multiplied by the meter reading, or 1.5 multiplied by 8%, which equals 12% (or as decimal fraction, 0.12) based on Equation 11.2.

$$\text{Actual Reading}_x = \text{CF}_x \times \text{Meter Display}_x \qquad (11.2)$$

The actual, corrected concentration of isobutyl acetate in this example would be 0.12 multiplied by 1.3%, which equals 0.156%. This is important from a couple of perspectives. First, it is accepted practice to have increased concern and caution, or an action level, when a CGM reading is ≥10% of the LFL[2]. In this example, the meter displayed 8% when, based on the correction factor, the actual reading was 12% of the LFL. While the measured 8% reading was less than an action level of 10%, the actual reading of 12% was greater and more representative of the hazardous conditions. From another perspective, relative to concentration, the 8% reading corresponded to a concentration of 0.104% or 1040 ppm (knowing that 1% equals 10,000 ppm). The actual concentration of isobutyl acetate, however, was substantially higher and equivalent to approximately 0.156% or 1560 ppm.

In many situations (if not most), more than one airborne flammable gas or vapor may be present or their presence is completely or relatively unknown. As a result, CFs are not always used. This implies, therefore, that there may be some degree of uncertainty regarding a CGM reading depending on the number and type of airborne flammable gases and vapors present at a given time in a given area. Accordingly, various guidelines are used to interpret CGM readings. As stated above, for example, ≥10% of the LFL has been accepted as an action level in some standard operating procedures. This signifies the need for extra caution or even withdrawal from the immediate area until the situation can be reassessed or better controlled.

CGMs are intended for use only in normal atmospheres, not atmospheres that are oxygen enriched or deficient. Oxygen concentrations substantially less than or greater than the normal 20.8% may cause erroneous readings. Accordingly, the level of airborne oxygen gas is usually measured simultaneously. Leaded gasoline vapors, halogens, silicone, and sulfur compounds can interfere with or damage the filament and decrease its sensitivity. Users must refer to the information associated with the specific make and model of instrument for precautions and limitations.

(ii) Oxygen Gas Indicator

The two principle components for operation of the oxygen indicator are the oxygen-sensing device and a meter readout. In some units, air is drawn into the oxygen detector with an aspirator bulb or pump, while in other units, equalization is allowed to occur between the ambient air and the sensor. An electrochemical sensor is used to determine the oxygen concentration in air. Components of a typical sensor include a counting and sensing electrode, a housing containing a basic electrolytic solution, and a semipermeable Teflon™ membrane.

Oxygen molecules diffuse through the membrane into the electrolytic solution. A minute electric current is produced by the reaction between the oxygen and the electrodes. This current is directly

[2] National Institute for Occupational Safety and Health, Occupational Safety and Health Administration, U.S. Coast Guard, Environmental Protection Agency, *Occupational Safety and Health Guidance Manual for Hazardous Waste Site Activities*, U.S. Government Printing Office, Washington, D.C., 2003.

proportional to the oxygen content of the sensor. The current passes through the electronic circuit, resulting in a direct-reading on the meter in units of percent. The meter is calibrated to read 0 to 10, 0 to 25, or 0 to 100% oxygen. When evaluating an unknown environment, the oxygen indicator on a combination CGM is essential to first determine if an oxygen-deficient or oxygen-enriched atmosphere is present.

UNIT 11 EXERCISE

OVERVIEW

The exercise will provide the fundamental concepts for conducting instantaneous monitoring of airborne combustible gases and vapors and oxygen gas using a combustible and oxygen gas meter. Place a check mark (✔) in the open box (☐) when you have obtained applicable material and completed the steps for the sampling method.

MATERIAL

1. Calibration of Combined Combustible and Oxygen Gas Survey Meter

- ☐ Combined combustible and oxygen gas meter
- ☐ Compressed reference or calibration gas (e.g., methane, pentane) for combustible gas and vapor
- ☐ Compressed reference or calibration gas for oxygen gas
- ☐ Regulators for compressed gas cylinders
- ☐ Appropriate flexible hose connecting meter to calibration gases
- ☐ Plastic air sampling bag to contain a volume of reference gas (optional depending on type of meter)

2. Sampling for Airborne Combustible Gas and Vapor or Oxygen Gas

- ☐ Combined combustible and oxygen gas meter
- ☐ Field monitoring data form (Figure 11.4)

METHOD

1. Pre- and Post-sampling Calibration of Combined Combustible and Oxygen Gas Survey Meter

- ☐ Refer to the survey meter manufacturer's instructions for assembly and operation of the instrument.
- ☐ Read the instructions that accompany the instrument or the compressed calibration gas for attaching the regulator to the gas cylinder. Assure that cylinders of compressed gases are secured so that they do not fall or roll off the bench top. Some manufacturers may require the transfer of calibration gas from a compressed gas cylinder to a plastic air sampling bag. In this case, the instrument is connected via a flexible hose to a port positioned on the air sampling bag that was already filled with calibration gas. Other instruments are connected via a flexible hose directly to the regulator on the compressed gas cylinder. Either setup permits efficient transfer of calibration gas to the meter.
- ☐ Follow the manufacturer's instructions for turning "ON" and subsequently connecting the instrument to the respective calibration gases.

Field Monitoring Data Form:
Real-Time Monitoring for Combustible Gases/Vapors (% LFL) and Oxygen (% O_2)
Using a Combined Combustible, Oxygen, and Toxic Gas/Vapor Meter

Facility Name and Location: ____________________

Monitoring Conducted By: ____________________

Date Monitoring Conducted: ____________________

Monitoring Instrument (Type/Manufacturer/Model): ____________________

Calibration Gas (LFL): ______________(___) **Calibration Gas (O_2):** ______________(___)

Calibration Gas (Toxic): ______________(___) **Correction Factor (LFL):**___________

Pre-Sampling Calibration: ____%LFL ____%O_2 ____ppm Toxic

Post-Sampling Calibration: ____%LFL ____%O_2 ____ppm Toxic

Air Temperature: ____ °C **Air Pressure:** ____mm Hg **Relative Humidity:** ____%

Identification of Area	Percent of LFL (%)	Oxygen (%)	Toxic Gas/Vapor (ppm)

Field Notes:

Figure 11.4 Field monitoring data form for instantaneous monitoring of combustible and oxygen gases.

- ☐ Slowly open the valve on the compressed gas cylinder or the plastic sampling bag filled with calibration gas to permit flow to the meter.
- ☐ Adjust the instrument according to the manufacturer's instructions so that readings for % LEL and % O_2 correspond to the concentrations of the respective reference gases. Close the valve on the gas cylinder or plastic air sampling bag and disconnect the instrument. The meter is ready for use.

2. Sampling for Airborne Combustible Gas and Vapor and Oxygen Gas

- ☐ Get approval to conduct monitoring in an area where airborne levels of combustible gas or vapor and oxygen gas are likely to be detectable and measurable and perform monitoring accordingly. Alternatively, conduct a simulated monitoring exercise including measurement of ambient levels of oxygen gas.
- ☐ Turn "ON" and zero the survey meter.
- ☐ Hold the meter approximately 4 ft above the floor or working surface and slowly survey the area.
- ☐ Read and record the measured concentration on the direct readout display.
- ☐ Complete a field monitoring data form, remembering to record your name, facility and location sampled, date, air temperature, air pressure, relative humidity, and combustible and oxygen gas meter manufacturer and model.

UNIT 12

Evaluation of Airborne Inorganic and Organic Gases and Vapors: Instantaneous Area Monitoring Using a Piston or Bellows Air Sampling Pump with a Solid Sorbent Detector Tube Medium

LEARNING OBJECTIVES

At the completion of Unit 12, including sufficient reading and studying of this and related reference material, learners will be able to correctly:

- Name, identify, and assemble the components of a common sampling train, including the specific sampling medium, used for conducting instantaneous monitoring of airborne inorganic and organic gases and vapors using solid sorbent detector tubes.
- Name, identify, and assemble the applicable calibration train that uses a primary standard for measuring flow rate of piston and bellows air sampling pumps.
- Calibrate a piston or bellows air sampling pump, with the applicable sampling medium in-line, using a manual or electronic frictionless bubble-tube.
- Summarize the principles of sample collection of airborne inorganic and organic gases and vapors using detector tubes.
- Conduct instantaneous monitoring of airborne inorganic and organic gases and vapors using detector tubes.
- Record all applicable calibration, sampling, and analytical data using calibration and field monitoring data forms.

OVERVIEW

It is frequently useful and necessary to instantaneously detect and measure the concentration of an airborne inorganic or organic gas or vapor. Devices are available that consist of a manual air sampling pump with detector tube media to serve this purpose. An air sample can be collected and within a minute or two, the presence of an individual contaminant or a class of contaminants can be determined qualitatively and quantitatively. Various detector tubes are available for numerous (>200) contaminants of concern.

Detector tubes may be used to screen areas to determine what areas should receive full shift monitoring via integrated monitoring techniques specific for the contaminant of concern. Detector tubes also may be used concurrently with full shift samples to trace sources of exposure and to track variations in exposure levels throughout the workshift.

DETECTOR TUBE MEDIA FOR GASES AND VAPORS

There are two classes of detector tubes — length of stain and colorimetric. In both cases, a tube is filled with a chemically coated solid sorbent that has been formulated to react specifically for the contaminant of interest (Figures 12.1 and 12.2). Common sorbents are silica gel and aluminium oxide that have been impregnated with an appropriate chemical reagent. The ends of the glass tube are sealed during manufacturing. When the ends of the tube are broken off and air containing the contaminant of interest is drawn through the tube, molecules of the gaseous contaminant contact the solid medium and react. Depending on the type of tube (i.e., length of stain or colorimetric), the reaction results in either a dark stain or color development. The length of stain or the intensity and shade of color development corresponds with the concentration of the gas or vapor contaminant in the sampled air. Length of stain tubes either have the concentration scale printed on them or the ratio of the length of stain to the total media length is compared against a chart of corresponding concentrations. When contaminated air is drawn through a colorimetric tube, the reagent-impregnated sorbent yields a progressive change in color intensity. At the completion of sampling, the sample tube is compared with a chart of tinted colors that correlate to the concentration of the gas or vapor present in the atmosphere. Most tubes are intended for collecting real-time data, but some tubes are available for integrated monitoring over an 8-h period.

PRECAUTIONS

Detector tubes are one of the easiest monitoring devices to use, but data must be interpreted with caution since the tubes are known to be relatively inaccurate. Accuracy is only ±25% in many cases and for some contaminants error is greater. Accuracy varies among different tubes. Therefore, a sample detector tube may reveal a concentration of a contaminant equal to 200 ppm, for example, but the actual concentration may be 150 to 250 ppm (200 ppm ± 25% or 200 ± 50 ppm). Variations in the grain size of the solid sorbent can cause channelling of the air sample through the tube. This can result in a stain demarcation that is not perpendicular to the walls of the tube, making it difficult to record a measurement.

Detector tubes have a limited shelf-life and should be refrigerated between uses. Detector tubes should be used only with the hand pump supplied by the manufacturer, as different manufacturer's pumps may have different flow rates relative to the volume of air required for the detector tubes. The flow rate of the air sampling pump is related to the absorption rate of the contaminant and for the chemical reactions to occur in the detector tubes.

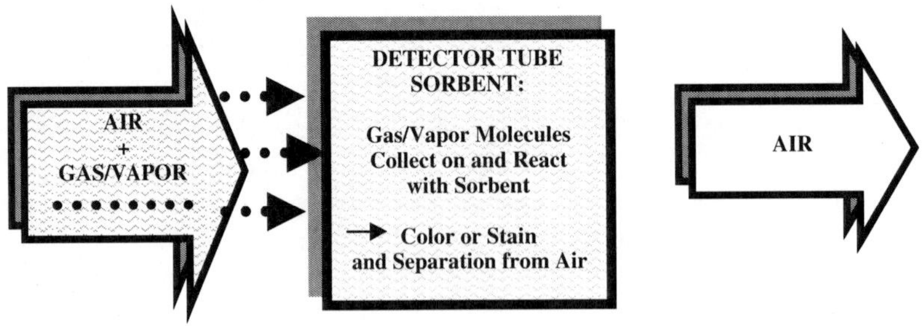

Figure 12.1 Sorption and reaction of gas/vapor in detector tube.

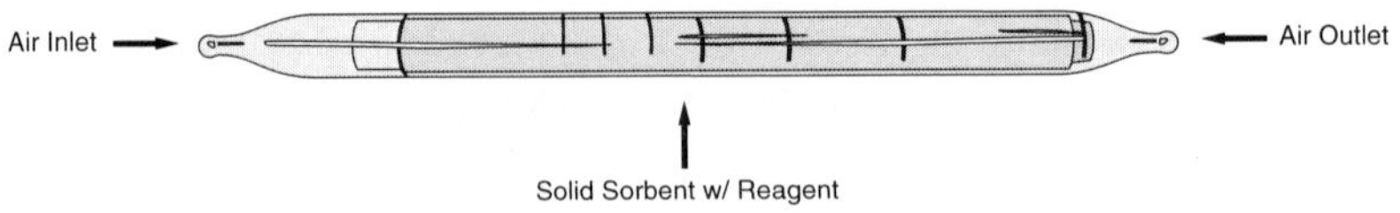

Figure 12.2 Standard glass detector tube filled with a solid sorbent coated with a chemical reagent.

SAMPLING AND ANALYSIS

Prior to sampling, a leak check is performed, as described in the Exercise section below, to assure that the pump is operating properly. Following the leak check, both ends of a glass detector tube are carefully snapped open for air to flow through during sample collection. (Some pumps have a built-in orifice to break off the end of the tubes.) The opened detector tube is inserted into the inlet orifice of the air sampling pump. The tubes often have an arrow on the side which should be pointed in the direction of the pump.

Two major types of air sampling pumps are available for use with detector tubes. Both types are manual and are referred to as piston pumps and bellows pumps (Figure 12.3). The piston pump is a manually operated device that draws a fixed volume of air through a detector tube with each stroke. A pump stroke refers to when the piston handle is withdrawn, causing evacuation of the pump cylinder. To assure that each pump stroke draws its full volume of air, a specified time period is required. Manufacturer's instructions specify the required time period for each pump.

The bellows pump is also manually operated and involves compression of a bellows structure. The compression of the bellows causes evacuation of the pump. Manual evacuation of either the piston pump or the bellows pump creates negative pressure within a pump chamber relative to atmospheric pressure, and accordingly, atmospheric air will surge into the pump. The air enters and passes through the detector tube prior to entering the evacuated pump. Once the pump is filled with sample air and pressure difference is zero, a known volume (e.g., 100 ml) of air has entered into and passed through the detector tube, that is, has been sampled. The volume sampled corresponds with the number of pump strokes or bellows compressions. Different tubes require different sample volumes, based on the number of strokes or compressions and depending on the contaminant and the concentration present.

Detectors yield a direct reading. Accordingly, laboratory analysis is unnecessary. A brief outlined summary of sample collection follows.

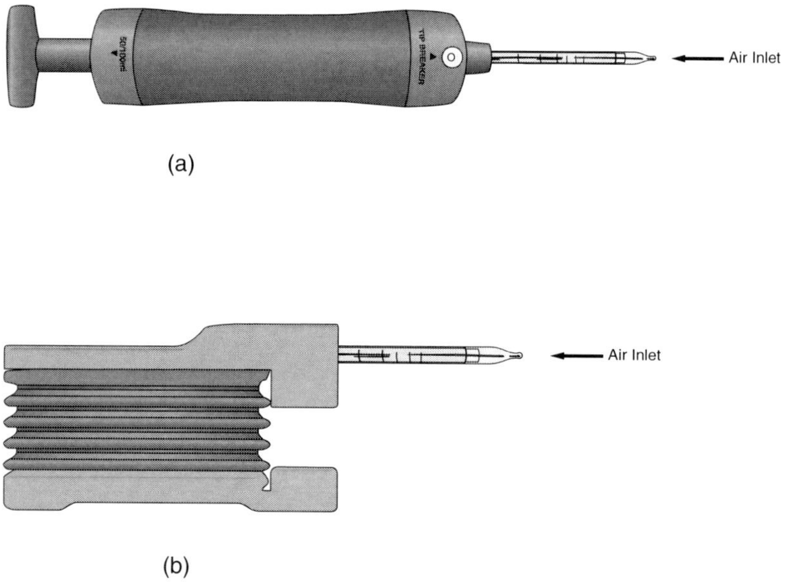

Figure 12.3 Sampling train for organic and inorganic gases and vapors consisting of either (a) a piston pump connected to a detector tube or (b) a bellows pump connected to a detector tube.

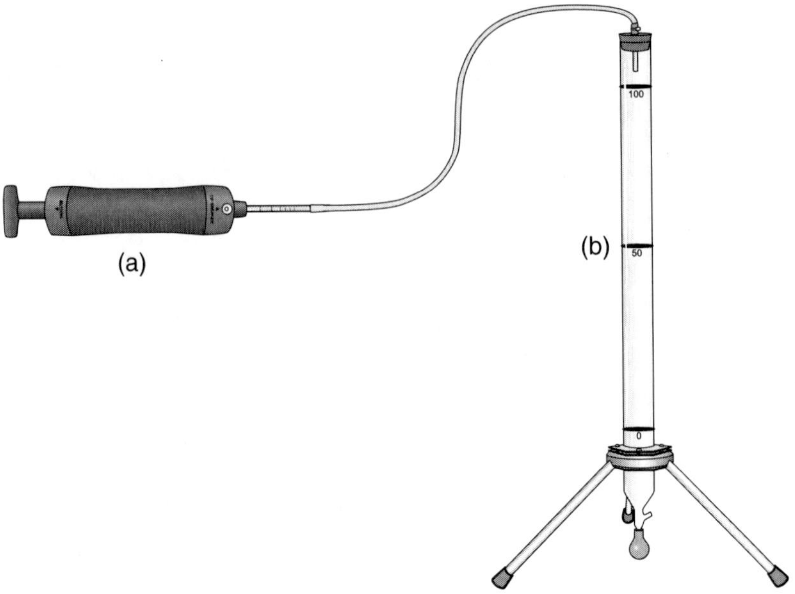

Figure 12.4 Calibration train for detector tube pump consisting of (a) a piston pump and detector tube in-line with (b) a frictionless bubble-tube.

1. Leak Test

- Manual air sampling pumps are checked to assure that there is no detectable leakage or malfunction.

2. Calibration

- A manual air sampling pump is calibrated periodically to adjust or determine flow rate using a manual or electronic calibrator with a glass detector tube in-line (Figure 12.4).
- Flow rates are determined by measuring the average time (T_{avg} [sec]) based on three trials ($\Sigma T_{1,2,3}/3$) for a bubble to traverse a 50 to 100 cm^3 volume (V [cm^3]) of the bubble-tube.

$$Q\ (cm^3/min) = \frac{V\ cm^3}{T_{avg}\ (sec)} \times \frac{60\ sec}{1\ min} \quad (12.1)$$

- A typical flow rate for sampling organic gases and vapors using solid sorbent detector tubes is 100 cm^3/min, based on a single pump stroke or compression.

3. Preparation for Monitoring

- Both ends of a labeled glass detector tube are snapped open.
- A sampling train is assembled and consists of a manual piston or bellows pump connected to a detector tube.

4. Conducting Monitoring

- The sampling train is held in a specific location for area monitoring.
- The piston or bellows pump is evacuated by manually withdrawing the piston or compressing the bellows, respectively.
- After specified time, usually 1 min, a pump stroke or compression is repeated and the detector tube is removed from the pump.
- The detection and measurement of the contaminant is determined based on the developed stain or color.

UNIT 12 EXERCISE

OVERVIEW

The exercise will provide the fundamental concepts for conducting instantaneous monitoring of an organic vapor using a piston or bellows pump with detector tube medium. Place a check mark (✔) in the open box (☐) when you have obtained applicable material and completed the steps for sampling.

MATERIAL

1. Calibration and Leak Test of Manual Air Sampling Pump

- ☐ Piston or bellows pump
- ☐ Solid sorbent detector tube
- ☐ 100-cm^3 manual or electronic frictionless bubble-tube
- ☐ 1-ft length $^1/_8$ in. internal diameter (i.d.) flexible hose
- ☐ Stopwatch (if manual calibrator is used)
- ☐ Soap solution
- ☐ Calibration data form (Figure 12.5)

2. Sampling for Gas or Vapor

- ☐ Applicable detector tube
- ☐ Detector tube end-breaker
- ☐ Plastic end caps (optional)
- ☐ Piston or bellows pump
- ☐ Field monitoring data form (Figure 12.6)

METHOD

1. Pre-sampling Leak Test of a Manual Air Sampling Pump

- ☐ Insert unopened detector tube into orifice of the piston or bellows pump.
- ☐ For the piston pump, withdraw the piston until it locks in the 100 cm^3 position, wait 2 min, and release the piston handle. The piston should retract back to the 0 cm^3 mark, indicating that there is no leakage.
- ☐ For the bellows pump, completely compress the bellows, release, and wait 10 min. The bellows should not have opened completely, indicating that there is no leakage.

Calibration Data Form:
Low-Flow, High-Flow, and Multi-Flow Air Sampling Pumps

Name and Location of Calibration: ______________________

Calibration Conducted By: ______________________ Date Calibration Conducted: ________

Calibration Instruments
(Type/Manufacturer/Model): ______________________

Air Sampling Pump
(Type/Manufacturer/Model): ______________________

Collection Medium In-Line: ______________ Calibrator Volume: ________cc

Air Temperature: ____°C Air Pressure: ____mm Hg Relative Humidity: ____%

Pump No.	Pre-Sampling Calibration Flow Rate (Q_{pre})					Post-Sampling Calibration Flow Rate (Q_{post})					Average Flow Rate
	Time (sec)				Q_{pre}	Time (sec)				Q_{post}	Q_{avg}
	T_1	T_2	T_3	T_{avg}	L/min	T_1	T_2	T_3	T_{avg}	L/min	L/min

Calibration Notes:

Figure 12.5 Calibration data form for piston or bellows air sampling pumps.

Field Monitoring Data Form:
Real-Time Monitoring for Inorganic and Organic Gases and Vapors
Using a Piston or Bellows Type Pump with Detector Tubes

Facility Name and Location: ____________________

Monitoring Conducted By: ____________________

Date Monitoring Conducted: ____________

Monitoring Instrument
(Type/Manufacturer/Model): ____________________

Pump Leak Check: ____ **Pass** **Contaminant:** ____________

Pump Calibration Check: ____**cc/min** **Detector Tube:** ____________

Instructions (Pump Stokes/Compressions):
____

Air Temperature: ____ **°C** **Air Pressure:** ____**mm Hg** **Relative Humidity:** ____**%**

Identification of Area	Concentration (ppm)

Field Notes:

Figure 12.6 Field monitoring data form for organic and inorganic gases and vapors.

2. Pre-sampling Calibration of a Piston Pump

- ☐ Complete a calibration form, remembering to record name, location, date, air temperature, air pressure, pump manufacturer and model, calibration apparatus, times for three trials, and average flow rate for calibration.
- ☐ Align a bubble to the 0 cm^3 mark of a frictionless bubble-tube. To do so, connect the flexible hose to the top of the burette and apply soap film to the bottom. Inhale lightly on the opposite (detached) end of the flexible tubing until the bubble reaches the 0 cm^3 mark.
- ☐ Break off the ends of a detector tube and insert into the inlet orifice of the piston or bellows pump.

- ☐ Connect the detached end of the 1/8 in. i.d. flexible hose to the inlet end of the detector tube.
- ☐ Withdraw the piston and lock it in the 100 cc position or compress the bellows. Time the period for the soap film to traverse 95 to 105 cm^3 and compare it to the pump manufacturer's recommended parameters.
- ☐ For the bellows pump, release and allow the bellows to expand. Assure that the volume traversed by the soap film was approximately 95 to 105 cm^3 following expansion of the bellows.

3. Air Sampling for Organic Vapor

- ☐ Get approval to conduct monitoring in an area where airborne levels of a measurable inorganic or organic vapor or gas are likely to be detectable and measurable and perform monitoring accordingly. Alternatively, conduct a simulated monitoring exercise using a known source of inorganic or organic gas or vapor provided by the instructor.
- ☐ Break off both ends of applicable detector tube and insert into the inlet orifice of the piston or bellows pump.
- ☐ Read the instructions for the specific detector tube to determine the number of pump strokes or compressions necessary for the contaminant.
- ☐ Align the index marks on the handle and back plate of the piston pump and pull the handle straight back to the sample volume specified for the contaminant/detector tube. The piston will lock in place. Wait the specified period of time for the evacuated pump to fill with air (more specifically, for the air to pass through the detector tube across the media). Rotate the handle 90° to release. Realign the index mark and repeat pump stroke procedures if specified.
- ☐ If a bellows pump is used, grip the pump between the thumb and the base of the forefingers and compress the bellows as far as possible and release. The end of the stroke is reached when the arrestor chain is fully tight. Repeat the process for the number of strokes indicated in specific detector tube instructions.
- ☐ Read and record the measurement of concentration.
- ☐ Complete a field monitoring data form, remembering to record your name, facility and location sampled, date, air temperature, air pressure, relative humidity, and piston or bellows pump and detector tube manufacturer and model.

UNIT 13

Evaluation of Airborne Toxic Gases and Vapors: Instantaneous Area Monitoring Using Organic Gas and Vapor Meters

LEARNING OBJECTIVES

At the completion of Unit 13, including sufficient reading and studying of this and related reference material, learners will be able to correctly:

- Name and identify the common instruments used for conducting instantaneous monitoring for airborne organic gases and vapors.
- Name, identify, and assemble the applicable calibration train that uses a known concentration of reference calibrants for calibrating photoionization detector (PID), flame ionization detector (FID), and infrared (IR) meters.
- Calibrate a PID, FID, or IR meter using an applicable known concentration of organic gas or vapor.
- Summarize the principles of sample collection of airborne organic gases and vapors using a PID, FID, or IR meter.
- Conduct instantaneous monitoring of airborne organic gases and vapors using a PID, FID, or IR meter.
- Record all applicable calibration and sampling data using field monitoring data forms.

OVERVIEW

Electronic devices are also available for instantaneous or real-time area monitoring of organic gases and vapors. These devices operate based on different principles, but they also share several similar characteristics. The devices consist of an air sampling pump that causes active flow of air and airborne contaminants into a chamber. The air and related organic gas and vapor molecules flow across a detector; this device qualitatively detects and generically identifies the contaminant as organic and quantitatively measures the concentration in parts per million or units on a direct readout display (Figure 13.1). In most cases, the identity of specific components of the airborne organic gas and vapor may be unknown or uncertain. Accordingly, data are commonly reported as total organic vapor (TOV).

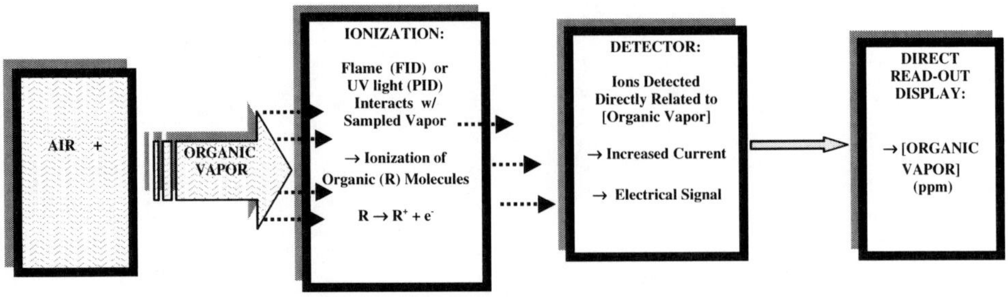

Figure 13.1 Schematic of toxic organic gas and vapor ionization meter.

MONITORING

(i) Flame Ionization Detector Meters

Meters with FIDs measure the concentration of total ionizable organic gases and vapors; if the identity of the gas or vapor is known, they also measure the concentration of specific organics in the atmosphere. Common components of a meter with an FID are a battery-operated air sampling pump, FID, cylinder of compressed hydrogen gas, readout, and sampling probe (Figure 13.2).

Response will vary depending upon the number of carbon atoms in the molecules. The response is not linear with the number of carbon atoms present. A gas or vapor is pumped into the instrument and passed through a hydrogen flame at the detector that ionizes the organic vapors with ionization potential less than 15 electronvolts (eV). Most organics have ionization potentials less than 12 eV. When most organic vapors burn, positively charged carbon-containing ions are produced and are collected by a negatively charged collecting electrode in the chamber. An electric field exists

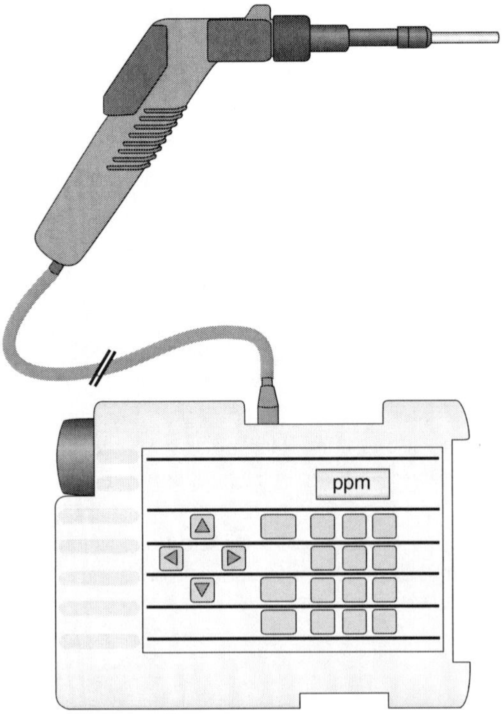

Figure 13.2 Combination FID/PID meter.

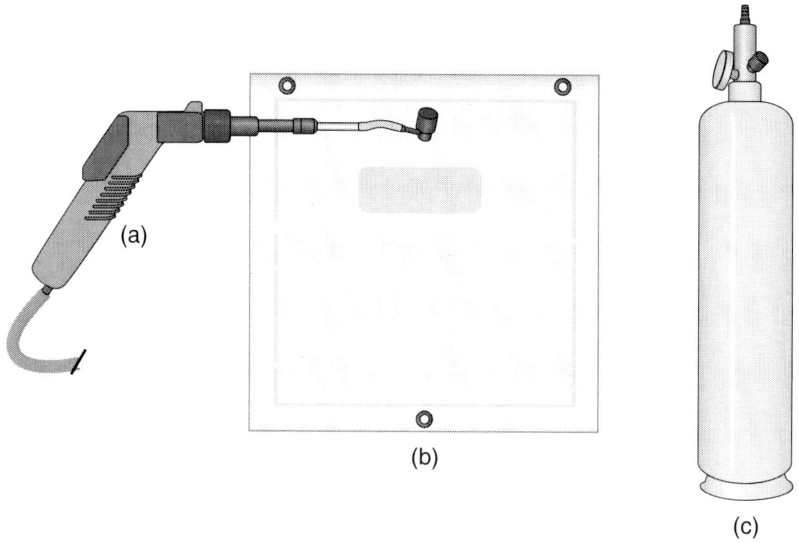

Figure 13.3 Calibration train for toxic organic gas and vapor meter consisting of (a) a combination FID/PID meter connected to (b) a sampling bag filled with reference gas from (c) a cylinder of compressed reference gas.

between the conductors surrounding the flame and a collecting electrode. As the positive ions are collected, a current proportional to the hydrocarbon concentration is generated on the input electrode. This current is measured with a preamplifier which has an output signal proportional to the ionization. Since a hydrogen flame is needed for combustion of sample, an adequate concentration of molecular oxygen must be present. FIDs are commonly calibrated using methane (e.g., 100 ppm) as a reference gas. For calibration, a sampling bag is commonly filled with reference gas obtained from a compressed gas cylinder. The sampling probe is then connected to the sampling bag (Figure 13.3).

(ii) Photoionization Detector Meters

Meters with PIDs also measure the concentration of total ionizable organic gases and vapors. Meters with a PID consist of a readout unit, a rechargeable battery, an amplifier, an ultraviolet (UV) lamp, a sensor ionization chamber, and a probe.

An electrical pump pulls the gas or vapor sample past a UV source. Constituents of a sample are ionized, producing an instrument response, if their ionization potential is equal to or less than the ionizing energy supplied by the UV lamp being utilized. The radiation produces positive ions of the contaminant molecules plus free electrons. The ions produce a current directly proportional to the number of ions produced, and in turn, directly proportional to the concentration of organic gas or vapor. The current is amplified, detected, and displayed on the meter. By varying the electron volts of UV light, a wide range of organic compounds can be detected and quantified, but not necessarily identified. UV lamps of 10.6 eV are relatively standard, but 10.2, 11.2, and 11.6 eV lamps are available. PIDs are commonly calibrated using isobutylene (e.g., 100 ppm) as a reference gas (see Figures 13.2 and 13.3).

(iii) Infrared Absorption Meters

Infrared spectrophotometric meters can detect and measure concentrations of specific organic gases and vapors, including carbon monoxide, carbon dioxide, anesthetic gases (e.g., nitrous oxide, halothane), sterilants (e.g., ethylene oxide), and fumigants (e.g., methyl bromide, ethylene dibromide). The instruments have a built-in sampling pump and a microprocessor that controls the meter and averages the electronic signal to yield direct readout. Compounds are detected based on absorption of a discrete wavelength of infrared light. In turn, absorbance values are proportional to the concentration of gas or vapor.

(iv) Portable Gas Chromatograph Meters

Portable gas chromatograph meters are versatile instruments for instantaneous monitoring. The instruments separate compounds and permit detection and quantification of specific organic compounds.

Sampled gas or vapor is passed through a column packed with a liquid phase on solid support or a solid phase support. Components of the sampled airstream separate as they pass through the column based on their inherent chemical characteristics. The individual components (analytes) are identified based on their retention time on the column and detected via a variety of commercially available detectors, typically FIDs. Laboratory models, as described in Unit 9, are capable of analyzing more complex molecules than portable gas chromatograph systems.

The organic gas and vapor meters are calibrated using known concentrations of reference gases. The reference gases are supplied via compressed gas cylinders. Known concentrations of vapors also can be prepared using solvents. Airborne concentrations of organic gases and vapors are measured in units of parts per million commonly known as TOV.

UNIT 13 EXERCISE

OVERVIEW

The exercise will provide the fundamental concepts for conducting instantaneous monitoring of airborne organic gases and vapors using an applicable electronic organic gas and vapor survey meter. Place a check mark (✔) in the open box (☐) when you have obtained applicable material and completed the steps for the sampling method.

MATERIAL

1. Calibration of Organic Gas and Vapor Survey Meter

- ☐ Organic gas and vapor meter
- ☐ Compressed organic reference or calibration gas
- ☐ Regulators for compressed gas cylinder
- ☐ Appropriate flexible hose for connecting meter to calibration gases
- ☐ Plastic air sampling bag to contain a volume of reference gas (optional depending on type of meter)

2. Sampling for Airborne Organic Gases and Vapors

- ☐ Organic gas and vapor meter
- ☐ Field monitoring data form (Figure 13.4)

Field Monitoring Data Form:
Real-Time Monitoring for Organic Gases and Vapors
Using Organic Gas/Vapor Meter

Facility Name and Location: ______________________________

Monitoring Conducted By: ______________________________

Date Monitoring Conducted: ____________

Monitoring Instrument (Type/Manufacturer/Model): ______________________________

FID Calibration Gas (Concentration): ____________________(____)
Pre-Sampling Calibration: ________ppm
Post-Sampling Calibration: ________ppm **Correction/Reduction Factor:** _____

PID Calibration Gas (Concentration): ____________________(____)
Pre-Sampling Calibration: ________ppm
Post-Sampling Calibration: ________ppm **Correction/Reduction Factor:** _____

Suspected Contaminant(s): ______________________________

Air Temperature: ____ °C **Air Pressure:** ____mm Hg **Relative Humidity:** ____%

Identification of Area	Concentration (ppm)	
	FID	PID

Field Notes:

Figure 13.4 Field monitoring data form for instantaneous monitoring of toxic organic gases and vapors.

METHOD

1. Pre- and Post-sampling Calibration of Organic Gas and Vapor Survey Meter

- ☐ Refer to the manufacturer's instructions for assembly and operation of the instrument.
- ☐ Read the instructions that accompany the instrument or the compressed calibration gas for attaching the regulator to the gas cylinder. Assure that cylinders of compressed gases are secured so that they do not fall or roll off the bench top. Some manufacturers may require the transfer of calibration gas from a compressed gas cylinder to a plastic air sampling bag. Alternatively, the gas can be transferred directly from the cylinder to the instrument provided a regulator is set at a specific flow rate. In this case, the instrument is connected via a flexible hose to a port positioned on the air sampling bag that was already filled with calibration gas. Other instruments are connected via a flexible hose directly to the regulator on the compressed gas cylinder. Either setup permits efficient transfer of calibration gas to the meter.
- ☐ Follow the manufacturer's instructions for connecting the instrument to the calibration gas cylinder or plastic air sampling bag.
- ☐ Turn "ON" the meter.
- ☐ Slowly open the valve on the compressed gas cylinder or the plastic sampling bag filled with calibration gas to permit flow into the meter.
- ☐ Adjust the instrument according to the manufacturer's instructions so that readings for the concentration of total organic gas or vapor correspond to the concentration of the reference gas or vapor. Close the valve on the gas cylinder or plastic air sampling bag and disconnect the instrument. The meter is ready for use.

2. Sampling for Airborne Organic Gases and Vapors

- ☐ Get approval to conduct monitoring in an area where airborne levels of organic gas or vapor are likely to be detectable and measurable and perform monitoring accordingly. Alternatively, conduct a simulated monitoring exercise using a known source of organic vapor provided by the instructor.
- ☐ Turn "ON" and zero the survey meter.
- ☐ Hold the meter at least 4 ft above the floor or working surface and slowly survey the area using a side-to-side sweeping motion.
- ☐ Read and record the concentration on the direct readout display.
- ☐ Complete a field monitoring data form, remembering to record your name, facility and location sampled, date, air temperature, air pressure, relative humidity, and organic gas and vapor meter manufacturer and model.

UNIT 14

Evaluation of Surface and Source Contaminants: Monitoring Using Wipe and Bulk Sample Techniques

LEARNING OBJECTIVES

At the completion of Unit 14, including sufficient reading and studying of this and related reference material, learners will be able to correctly:

- Name and identify the common supplies and instruments used for conducting wipe and bulk sampling.
- Summarize the principles of monitoring for surface and source contaminants.
- Conduct wipe and bulk sampling for surface and source contaminants.
- Record all applicable sampling data using field monitoring data forms.

OVERVIEW

Physical, chemical, or biological agents may deposit and accumulate on various surfaces. These deposits may result from direct spillage and via settling of airborne particulate. Contaminated surfaces are potentially a significant source of external exposure due to direct contact with skin and subsequent indirect contact with eyes due to contaminated hands. Surface contamination also can pose an ingestion hazard due to indirect transfer of contaminants from skin (i.e., hands and fingers), ingesta (e.g., food, chewing gum), and tobacco products (e.g., cigarettes, chewing tobacco) placed into the mouth. In addition, eventual resuspension of settled particulate can generate secondary dusts that may enter the body via inhalation.

Because of the potential for surface contamination, inanimate surfaces in many types of occupational environments need to be monitored periodically to determine if contaminants are detectable, identifiable, and measurable. These surfaces include bench tops, lockers, cafeteria tables and chairs, and personal protective equipment (PPE). In relation, since the dermal surface can become contaminated, the skin is often a site of sample collection. The effectiveness of dermal and respiratory PPE, such as gloves and respirators, can be evaluated in part by collecting samples from the interior surface of the PPE and the surface of the underlying skin. Surface sampling also can be conducted to evaluate the efficiency of equipment decontamination procedures.

Bulk samples are often necessary to determine the identity of a given contaminant that may be a component of raw materials, by-products, products, building materials, and even the air. Manufacturing grade liquid petroleum-based solvents are frequently a mixture of compounds. Even if information regarding the type of components is available, there may remain uncertainty regarding the percentage of a given constituent. As a result, it is often necessary to obtain a bulk sample of a liquid solvent to qualitatively and quantitatively identify its chemical composition. Solid powders, such as pigments, used to manufacture colorants and paints may contain a variety of solids and impurities that include chromium, lead, and crystalline silica. The qualitative and quantitative chemical composition will provide information relative to the composition of airborne dusts. Building materials are often sampled since many may contain toxic agents. Perhaps most notable is the presence of asbestos in insulation and lead found in surfaces coated with lead-based paint. In addition, bulk or grab samples of air are often necessary to identify a spectrum of toxic substances that may warrant additional monitoring and compounds that may interfere with a given monitoring or analytical method.

SAMPLING

Surface sampling literally involves wiping or swabbing a surface suspected of contamination and subsequently submitting the sample for laboratory analysis. Usually, a 100 cm^2 surface should be wiped or swabbed for chemical agents and even less (e.g., 1 cm^2) for microbiological agents. Two major supplies required for surface sampling are glass fiber and paper filters for wipe sampling and cotton and foam swabs for swab sampling (Figure 14.1). In addition, disposable latex or vinyl gloves should be worn during sampling. Paper filters are typically used for wipe samples for metals and glass fiber filters for sampling organic compounds. Individual filters are inserted and stored in clean capped vials prior to use and are returned to the vials after wipe sampling is completed. Depending on the agent, filters are used either dry or moistened with deionized water or other solvent to facilitate sample collection. Cotton swabs are most applicable for surface sampling of microbiologic agents such as bacteria and fungi. The swabs are immersed in a buffer solution within a vial that is sterilized prior to use. Following a swabbing procedure, the swab is returned to the vial containing sterile buffer.

Bulk sampling of liquids and solids are typically very simplistic. Rigid containers, such as jars, are generally used to collect and deposit liquid samples and non-rigid sealable plastic bags are typically used for solid samples. Straight hollow glass tubes called thief tubes and coliwassa tubes

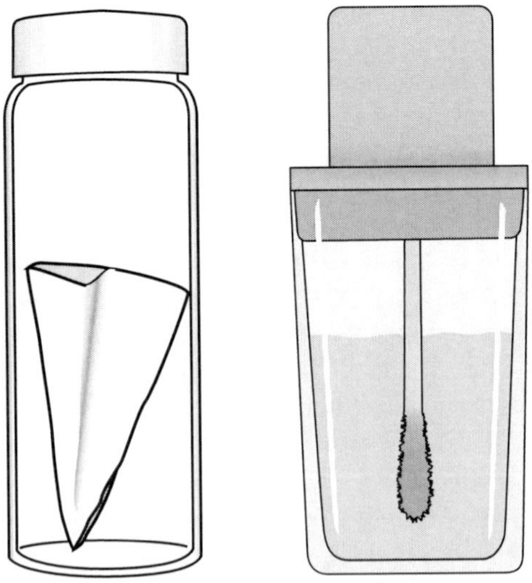

(a) (b)

Figure 14.1 Sampling medium for surface sampling consists of either (a) a filter with storage vial for wipe sampling or (b) a cotton swab immersed in buffer for swab sampling.

are frequently used to collect otherwise relatively inaccessible liquid samples from drums and tanks. The sample is then deposited in a jar and capped. Solid materials, such as powders, can be collected directly from the source; devices such as corers and scoops can also be used. Intact solids, such as building materials, can be sampled by simply placing a representative section in a plastic bag or using a small corer to collect a plug sample. Large corers, augers, and shovels are often used to collect bulk samples of soil. Bulk samples for air are slightly more complex and commonly involve the use of a low- or high-flow air sampling pump connected to a special plastic air sampling bag. Air is pumped from the atmosphere being sampled into the bag. Also, small rigid stainless cylinders or canisters can be pressurized with sampled air. In addition, high-volume bulk air samples of particulates can be collected on overloaded filter media and gaseous agents on high capacity sorbent tubes. Some examples are shown in Figures 14.2 and 14.3.

ANALYSIS

Analyses of surface and bulk samples are dependent on the type of contaminants that need to be detected and measured. Wipe samples for metals using paper filters are analyzed using atomic absorption or inductively coupled plasma spectrophotometry (refer to Unit 7). Samples for surface organics using glass fiber filters and airborne organics using sampling bags are analyzed using methods such as gas chromatography (GC) (refer to Unit 9). Swab samples for microorganisms using sterile cotton swabs are analyzed by culturing on a nutrient agar, incubating, and subsequently counting colonies (refer to Unit 15).

Bulk samples of liquids; solids; and air for metals, organics, and microbes are analyzed using the same methods specified previously. Some liquid samples may be analyzed for certain organics using ultraviolet or visible (UV/Vis) light spectrophotometry (refer to Unit 10). Polarizing light microscopy is commonly used to analyze bulk samples of building materials for asbestos. X-ray diffraction is a common analytical instrument used for analyzing bulk samples of powders for crystalline silica.

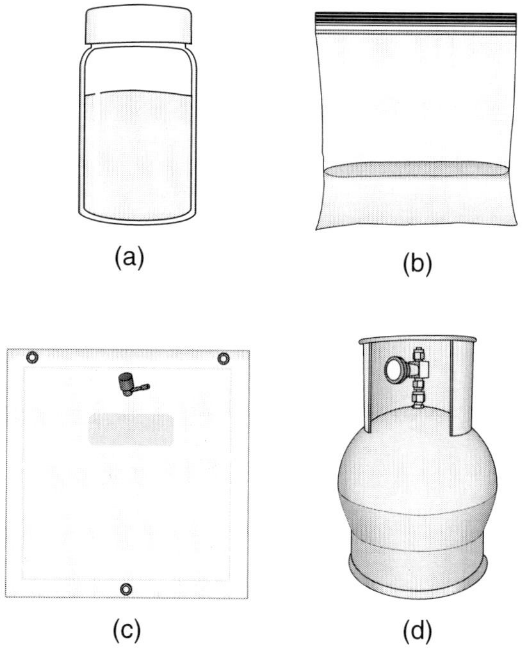

Figure 14.2 Containers for collecting bulk samples consist of (a) rigid jars for liquid or solid samples, (b) nonrigid sealable plastic bags for liquid or solid samples, (c) nonrigid plastic air sampling bags for gaseous chemicals, and (d) rigid stainless steel canister for air sampling gaseous chemicals.

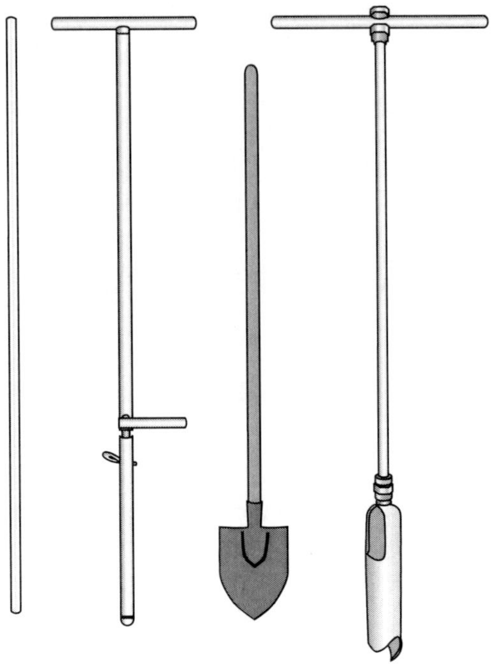

Figure 14.3 Devices for collecting bulk samples consist of (a) a hollow glass tube or thief for sampling water and other liquids, (b) a corer for sampling soil, (c) a shovel for sampling soils and other solids, and (d) an auger for sampling soils.

UNIT 14 EXERCISE 1

OVERVIEW

The exercise will provide the fundamental concepts for conducting monitoring for surface contamination of lead dust using a paper filter wipe sampling technique. The reader is referred to Unit 7 for a related analytical method using atomic absorption spectrophotometry. Place a check mark (✔) in the open box (☐) when you have obtained applicable material and completed the steps for the sampling method.

MATERIAL

1. Sampling for Surface Contamination

☐ Clean, labeled vial with cap
☐ Filter paper
☐ Deionized water
☐ Disposable latex gloves
☐ Field monitoring data form (Figure 14.4)

Field Monitoring Data Form:
Surface and Bulk Sampling

Facility Name and Location: ______________________________

Monitoring Conducted By: ______________________________

Date Monitoring Conducted: ____________

Monitoring Instrument(s) (Type/Manufacturer/Model) and Supplies: ______________

____

Area	Sample Number	Wipe, Swab or Bulk Sample Type	Suspected Contaminant(s)

Field Notes:

Figure 14.4 Field monitoring data form for surface and bulk sampling.

METHOD

1. Sampling for Surface Contamination

- ☐ While wearing disposable latex gloves, fold and place a new filter paper in a clean, labeled vial and cap. Prepare several vials containing filters.
- ☐ Get permission to conduct surface sampling in an area where there is a possibility of work surfaces contaminated with lead dust. Alternatively, conduct a simulated surface sampling activity.
- ☐ When ready to begin, assure that clean disposable latex gloves are worn.
- ☐ Open a labeled vial and withdraw and unfold a filter.
- ☐ Moisten the filter with deionized water.
- ☐ Wipe a surface area of approximately 100 cm^2.
- ☐ Fold the used filter with the exposed side inward and insert it back into the original vial.
- ☐ Complete a field monitoring data form including name, location, and area sampled.
- ☐ Submit the sample for laboratory analysis for total lead.
- ☐ At least one blank filter handled with the same technique, but without sampling, should be submitted for each sampled area.

UNIT 14 EXERCISE 2

OVERVIEW

The exercise will provide the fundamental concepts for conducting monitoring for source contamination for organic components of a solvent using a liquid bulk sampling technique. The reader is referred to Unit 9 for a related analytical method using gas chromatography. Place a check mark (✔) in the open box (☐) when you have obtained applicable material and completed the steps for the sampling method.

MATERIAL

1. Bulk Sampling of Liquid Organic Solvent

- ☐ Clean, labeled jar with cap
- ☐ Hollow glass tube (e.g., thief or coliwassa tube)
- ☐ Pipette suction bulb
- ☐ Appropriate goggles and impermeable synthetic rubber gloves
- ☐ Field monitoring data form

METHOD

1. Bulk Sampling of Liquid Organic Solvent

- ☐ Get permission to collect a liquid sample from a 55-gal drum containing a known industrial solvent. Alternatively, conduct a simulated bulk sampling activity using a clean drum filled with water.
- ☐ When ready to begin, assure that goggles and impermeable gloves are worn.
- ☐ Place the pipette suction bulb over the top open end of the glass tube.

- ☐ Compress the pipette suction bulb and insert the glass tube into the drum and below the liquid surface.
- ☐ Allow the pipette suction bulb to expand and then slowly remove the glass tube from the drum.
- ☐ Place the open bottom end of the glass tube into a sample collection jar and compress the pipette suction bulb to dispense the liquid sample into the jar.
- ☐ Place the cover on the labeled jar.
- ☐ Complete a field monitoring data form, remembering to record name, location, and area sampled.
- ☐ Submit the sample for laboratory analysis for profile of generic or specific organic components.

UNIT 15

Evaluation of Airborne Bioaerosols: Integrated Area Monitoring Using an Air Sampling Pump with an Impactor and Nutrient Agar Medium

LEARNING OBJECTIVES

At the completion of Unit 15, including sufficient reading and studying of this and related reference material, learners will be able to correctly:

- Name, identify, and assemble the components of a common sampling train, including the specific sampling medium, used for conducting integrated monitoring of airborne bioaerosols.
- Name, identify, and assemble the applicable calibration train that uses a primary standard for measuring flow rate of a high-flow air sampling pump.
- Calibrate a high-flow air sampling pump, with the applicable sampling medium in-line, using a manual or electronic frictionless bubble-tube.
- Summarize the principles of sample collection of airborne bioaerosols.
- Conduct integrated monitoring of airborne bioaerosols.
- Prepare samples for analysis of collected biological agents using incubation.
- Summarize the principles of sample analysis of agar used for collection of airborne bioaerosols.
- Conduct analysis of a sample for bioaerosols.
- Perform applicable calculations and conversions related to pre- and post-sampling flow rate of an air sampling pump, average flow rate of an air sampling pump, sampling time, sampled air volume, number of colonies counted, and concentration of bioaerosols in units of colony forming units per cubic meter (CFU/m^3).
- Record all applicable calibration, sampling, and analytical data using calibration, field monitoring, and laboratory analysis data forms.

OVERVIEW

Bioaerosols represent a group of viable or living particulates, including bacteria, fungi, and protozoa. Although specific occupational exposure limits are not established for airborne viable organisms, monitoring is often conducted to determine if airborne organisms, as bioaerosols, can be detected and measured for comparison to general recommended guidelines and levels in background control areas. The organisms are frequently combined with airborne mists, sprays, and dusts; the particle size distribution can be classified as nonrespirable and respirable fractions.

The collection of airborne bioaerosols can be achieved through use of various methods, including filtration and impingement. A relatively accurate method involves the use of specialized sieve-

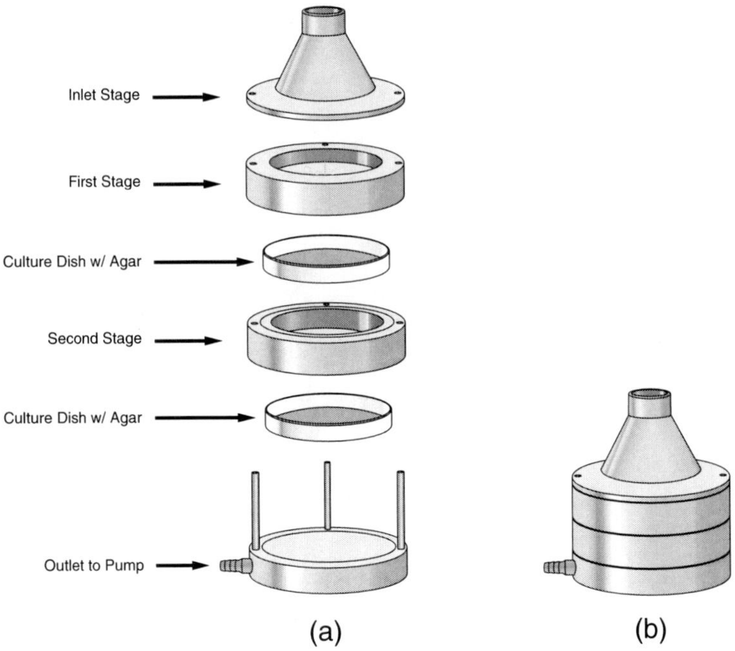

Figure 15.1 Two-stage impactor with culture dishes containing the agar medium. (a) Exploded view and (b) assembled view. The upper or first stage has larger diameter holes than the lower or second stage.

like devices called impactors (Figure 15.1). To detect the presence of organisms, they are collected on a medium containing agar (a solidifying agent) and nutrients, allowing the growth of colonies of microorganisms that are visible for counting. Dishes of a specific type of agar are positioned within the impactor during monitoring so that the bioaerosols are initially separated based on particle sizes by the seives and deposited on respective dishes via impaction on solid agar for growth, detection, and measurement.

GROWTH MEDIA FOR BIOAEROSOLS

Agar serves as growth media for numerous microorganisms such as bacteria and fungi. Commonly used media are trypticase soy agar (TSA) for bacteria and for fungi malt extract agar (MEA) or potato dextrose agar (PDA). For other organisms or for more selective monitoring of specific organisms, more specific growth media can be used, depending on the type of microbial monitoring that is conducted.

The plates (dishes) of agar media may be laboratory-prepared or purchased ready-made from commercial suppliers.

If laboratory-prepared, a commercially available dehydrated medium, deionized water, and sterile petri dishes (culture plates) are needed. Use plastic petri dishes that are a nominal 100 mm diameter and 15 mm deep. Do not use glass petri dishes, since they are a breakage hazard and many of these are too deep to fit into the single- or two-stage impactors. Instructions for the next steps are listed below:

- Determine the number of culture plates needed for the sampling (e.g., using a two-stage impactor, two per sample + two for field blanks + two for the calibration process).
- Allow for 17 to 20 ml medium per plate.
- Follow the dehydrated medium manufacturer's instructions for medium preparation, using the appropriate weight of medium for the volume to be made.

- Sterilize the medium in a steam autoclave at the appropriate temperature and time (will usually be 15 min at 121°C).
- Allow the autoclaved medium to cool to approximately 60°C (preferably in a water bath).
- Pour approximately 17 to 20 ml medium in to each plate and allow each to solidify without moving.

So that agar surface is free of excess moisture, it is advisable to prepare the plates at least 24 h before use. Prepared plates can be stored in plastic sleeves for several weeks.

If purchasing commercially prepared plated medium, use the nominal 100 mm diameter, 15 mm deep plates. These will contain 15 to 17 ml medium. Some manufacturers have double fill plates of fungal isolation media that contain 35 to 40 ml medium. If purchasing plates of fungal growth media, do not use these double fill plates since this is too much agar and may block the holes in the impactor.

MONITORING

Monitoring using an impactor involves preparation of the sampling medium: in this case, nutrient agar in culture dishes positioned within the impactor. Several types of impactors are available. Common impactors consist of one, two, or six stages. A culture plate containing agar is positioned within each stage for actual impaction and collection of the bioaerosols.

A single-stage impactor may be used if knowledge of respirable and nonrespirable particle sizes is not needed. A two-stage cascade impactor can separate respirable from nonrespirable fractions. A six-stage cascade impactor permits even more specific separation based on particle sizes. Regardless of the number of stages, an ultrahigh-flow air sampling pump is needed to collect the air sample. The typical air sampling train consists of an air sampling pump connected to the impactor with 1/4 or 1/2 in. internal diameter (i.d.) heavy gauge flexible hose (Figure 15.2). The sampling train is commonly operated at a flow rate of 1 ft^3/min which is equivalent to 28.3 l/min. A minimum air volume of 0.1 m^3 should be collected depending on the conditions. Prior to and in between use, the cascade impactor should be disinfected using a chemical disinfectant (e.g., isopropanol) or sterilized using an autoclave to minimize cross-contamination of samples.

A brief outlined summary of sample collection follows.

(i) Calibration

- An ultrahigh-flow air sampling pump is calibrated pre- and post-sampling to adjust or determine flow rate using a calibrator with a capacity of measuring flow rates up to 30 l/min. The ultrahigh-flow pumps frequently have a rotameter calibrated by the manufacturer. The flow rate can also be checked and calibration conducted using an electronic bubble- or dry-tube type calibrator as shown in Figure 15.3 using a homemade calibration chamber/jar.
- Pre- (Q_{pre}) and post-sampling (Q_{post}) flow rates are determined by measuring the average time (T_{avg} sec) based on three trials ($\sum T_{1,2,3}/3$) for a bubble to traverse a 30-l volume (V [l]) of the bubble-tube.

$$Q\,(\mathrm{l/min}) = \frac{\mathrm{V\,(l)}}{T_{avg}\,(\mathrm{min})} \tag{15.1}$$

- The average flow rate (Q_{avg}) is based on the average of pre- (Q_{pre}) and post-sampling (Q_{post}) flow rates.

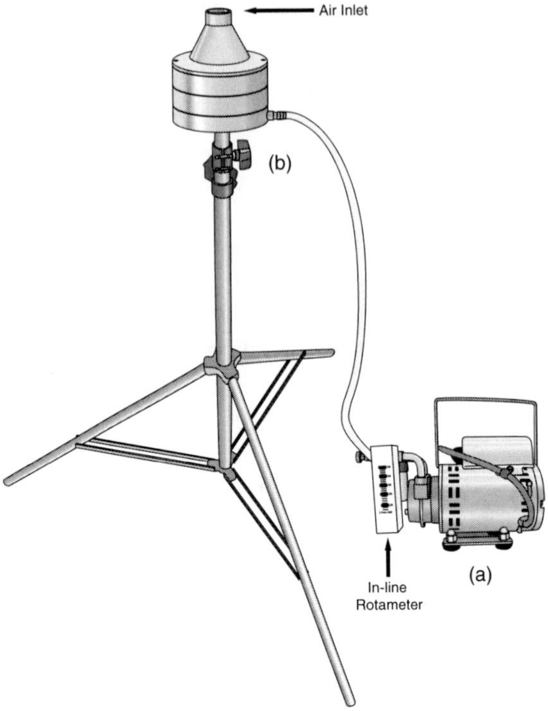

Figure 15.2 Sampling train for bioaerosols composed of (a) an ultrahigh-flow vacuum pump connected with flexible hose to (b) a two-stage impactor with culture dishes containing the agar medium.

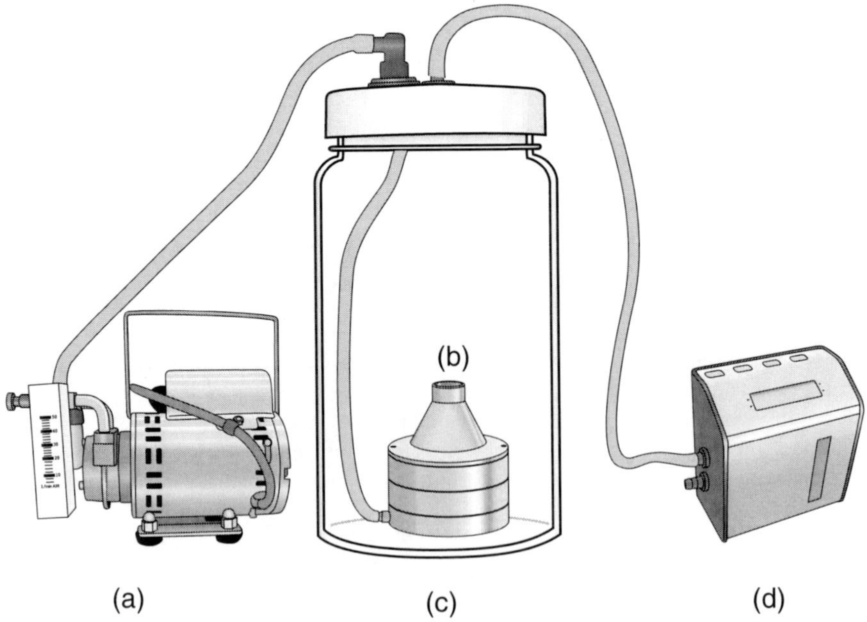

Figure 15.3 Calibration train for bioaerosols composed of (a) an ultrahigh-flow vacuum pump connected with flexible hose to (b) a two-stage impactor with culture dishes containing the agar medium within (c) a calibration jar in-line with (d) a high-volume electronic frictionless bubble-tube.

$$Q_{avg} = \frac{Q_{pre} + Q_{post}}{2} \qquad (15.2)$$

- Typical flow rate for sampling bioaerosols using a two- or six-stage cascade impactor is 28.3 l/min (1 ft^3/min).

(ii) Preparation for Monitoring

- Sterilized petri dishes and agar are labeled and positioned within a disinfected impactor.
- A sampling train is assembled and consists of ultrahigh-flow pump, 1/4- or 1/2-in. i.d. heavy gauge flexible hose, and impactor that contains culture plates containing agar.

(iii) Conducting Monitoring

- The pump is turned "ON" and start time is recorded.
- After a specified time, the pump is turned "OFF" and stop time is recorded.
- Culture dishes containing agar are removed from the cascade impactor, covered, and transported to laboratory for incubation and subsequent analysis.

ANALYSIS

Analysis involves an incubation period to allow for the growth of detectable and measurable colonies of microorganisms (Figure 15.4). The incubation time, temperature, humidity, and lighting varies depending on the type of organisms that were collected and are being selected for analysis. Following a prescribed incubation, the culture dishes are observed for detectable and countable microbial colonies. Dishes can be positioned under a colony counter — an illuminated grid surface with a magnification lens — to facilitate detection and counting. Methods, such as subculturing, staining and microscopy, polymerase chain reaction (PCR), and immunoassay, can be used to identify the specific genera and species of organisms collected and present in the colonies.

(i) Determination of Concentration of Bioaerosols

- The sampled volume of air is determined by multiplying sampling time (T [min]) by average sampling flow rate (Q [l/min]) and converting liters (l) to cubic meters (m^3).

$$Air\ \text{Volume}\ (m^3) = T\,(\text{min}) \times Q_{avg}\,(\text{l / min}) \times \frac{1\ m^3}{10^3\ \text{l}} \qquad (15.3)$$

- The concentration of microorganisms is determined based on the number of colonies counted divided by the volume of air sampled. The assumption is made that each visible colony has arisen from one bacterial cell or from one fungal spore or fragment of fungal hypha, but this

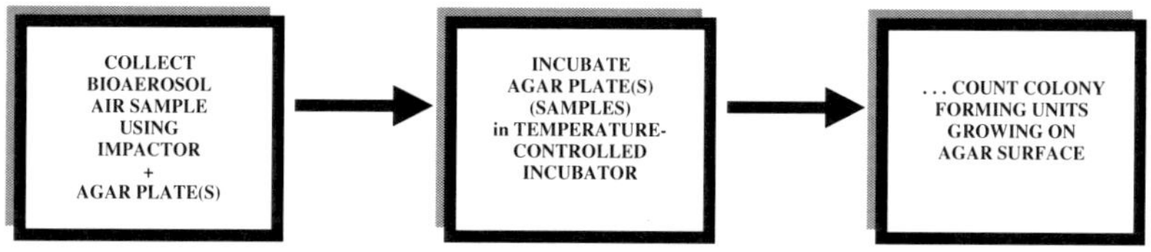

Figure 15.4 Stages of bioaerosol sample collection and analysis.

may not always be the case, so the concept of colony-forming units is used. Concentration is commonly expressed in units of colony forming units per cubic meter of air (CFU/m^3).

$$\text{Bioaerosol (CFU / m}^3\text{)} = \frac{\textit{Colonies}}{\text{Volume Sampled (m}^3\text{)}} \tag{15.4}$$

UNIT 15 EXERCISE

OVERVIEW

The exercise will provide the fundamental concepts for conducting integrated monitoring for bioaerosols using an impactor and agar growth medium. Although a two-stage impactor is used here, other impactors (i.e., one- or six-stage) can be used with minimal modification to the exercise. In addition, a common and related analytical method using incubation and a colony counter is introduced. The exercise will focus on sampling and analysis of airborne bacteria. Place a check mark (✔) in the open box (☐) when you have obtained applicable material and completed the steps for sampling and analytical methods.

MATERIAL

1. Calibration of Air Sampling Pump

- ☐ Ultrahigh-flow air sampling pump
- ☐ Representative sampling medium consisting of a two-stage impactor containing culture dishes filled with agar
- ☐ Two 2-ft or longer lengths of 1/4- or 1/2-in. i.d. heavy gauge flexible hose
- ☐ 4-l glass calibration jar (if not available, calibrator may be connected to the cone-shaped cover of the impactor with an appropriately sized, air-tight, stopper)
- ☐ High-volume (30 l) electronic frictionless bubble-tube calibrator
- ☐ Calibration data form (Figure 15.5)

2. Sampling for Bioaerosol

- ☐ Ultrahigh-flow air sampling pump
- ☐ One 2-ft or longer length of 1/4- or 1/2-in. i.d. heavy gauge flexible hose
- ☐ Two-stage impactor
- ☐ Two culture dishes containing 17 to 20 ml sterilized trypticase soy agar
- ☐ Field monitoring data form (Figure 15.6)

3. Analysis of Sample for Bioaerosols

- ☐ Incubator
- ☐ Colony counter (preferable, but not essential)
- ☐ Laboratory analysis data form (Figure 15.7)

Calibration Data Form:
Low-Flow, High-Flow, and Multi-Flow Air Sampling Pumps

Name and Location of Calibration: ____________________

Calibration Conducted By: ____________________ **Date Calibration Conducted:** ________

Calibration Instruments (Type/Manufacturer/Model): ____________________

Air Sampling Pump (Type/Manufacturer/Model): ____________________

Collection Medium In-Line: ____________ **Calibrator Volume:** ____cc

Air Temperature: ____°C **Air Pressure:** ____mm Hg **Relative Humidity:** ____%

Pump No.	Pre-Sampling Calibration Flow Rate (Q_{pre})					Post-Sampling Calibration Flow Rate (Q_{post})					Average Flow Rate
	Time (sec)				Q_{pre}	Time (sec)				Q_{post}	Q_{avg}
	T_1	T_2	T_3	T_{avg}	L/min	T_1	T_2	T_3	T_{avg}	L/min	L/min

Calibration Notes:

Figure 15.5 Calibration data form for high-flow vacuum air sampling pump.

Field Monitoring Data Form:
Integrated Monitoring for Bioaerosols

Facility Name and Location: ________________ **Contaminant Sampled:** ☐ **Bacteria** ________
☐ **Fungi** ________

Monitoring Conducted By: ________________

Collection Medium: ________________ **Date Monitoring Conducted:** ________________

Method(s)(Source/Number/Name): ________________

Monitoring Instruments (Type/Manufacturer/Model): ________________

Air Temperature: ____°C **Air Pressure:** ____mm Hg **Relative Humidity:** ____%

Identification of Personnel or Area	Field Sample	Pump No.	Pre-Sample Calibration Flow Rate	Post-Sample Calibration Flow Rate	Average Flow Rate (Q)	Sample Start-Time	Sample Stop-Time	Total Sample Time (T)	Volume Sampled (Vol)

Field Notes:

Figure 15.6 Field monitoring data form for bioaerosols.

Laboratory Analysis Data Form:
Analysis of Bioaerosol Samples for Microbial Colonies

Monitored Facility Name and Location: ____________________ **Contaminant Sampled:** ____________

Monitoring Conducted By: ____________________ **Date Monitoring Conducted:** ____________

Laboratory Facility Name and Location: ____________________ **Contaminant Analyzed:** ____________

Analysis Conducted By: ____________________ **Date Analysis Conducted:** ____________

Nutrient Agar Collection Medium (Type/Manufacturer): ____________

Colony Counter (Type/Manufacturer/Model):

Temperature: ____ °C **Air Pressure:** ____mm Hg **Relative Humidity:** ____%

Laboratory Sample Identification No.	Field Sample Identification No.	Incubation Temperature (°C)	Incubation Time (Hours)	Number Colony Forming Units (CFUs)	Average Flow Rate (Q) (l/min)	Total Sample Time (T) (min)	Volume Sampled (Vol) (m^3)	Conc. (CFU/m^3)

Laboratory Notes:

Figure 15.7 Laboratory analysis data form for analysis of bioaerosols.

METHOD

1. Pre-sampling Calibration of Air Sampling Pump

- ☐ Plug pump into AC outlet, turn "ON," and allow it to operate for 5 min prior to calibrating.
- ☐ Complete a calibration data form, remembering to record name, location, date, air temperature, air pressure, pump manufacturer and model, calibration apparatus, times for three trials, and average flow rate for pre- and post-sampling calibrations.
- ☐ Obtain representative sampling medium for use during calibration of the air sampling pump. This should consist of a two-stage impactor containing two culture dishes containing agar. These two dishes should not be used for the actual sampling, but may be reused in the impactor for the post-sampling calibration check. After being used for the calibration checks, these dishes of medium may be discarded.
- ☐ Connect one end of the flexible hose from inside the calibration jar to the hose connection on the two-stage impactor.
- ☐ Place the impactor in the calibration jar and tighten the lid securely. Wrap plastic tape around the perimeter of the sealed lid to reduce leakage of outside air.
- ☐ Connect the end of a flexible hose to the air sampling pump.
- ☐ Connect the other end of the same hose to the tube projecting out of the lid of the calibration jar so that there is a connection between the impactor to the lid tube and lid tube to pump via flexible hoses.
- ☐ Connect the end of the second flexible hose to the second tube that projects from the lid of the calibration jar.
- ☐ Connect the remaining open end of flexible hose to the electronic frictionless bubble-tube; this constitutes the calibration train.
- ☐ Aspirate the bubble solution into the calibration device to form a soap film (bubble).
- ☐ Record the time (T [sec]) it takes the soap film (bubble) to traverse a volume of 30 l.
- ☐ If the flow rate is not 28.3 l/min, adjust the flow control on the pump and repeat calibration procedure until desired flow rate is achieved.
- ☐ Once the desired flow rate is achieved (in this example 28.3 l/min), repeat the calibration check two to three times and calculate the average pre-sampling flow rate.
- ☐ Disconnect the pump from the calibration remainder of the calibration apparatus and turn "OFF." Calibrated air sampling pump is ready for use.

2. Sampling for Airborne Bioaerosols

Note: The following is described for bacteria. Should the sampling be for fungi, a suitable fungal medium should be used (e.g., PDA). The procedure would be the same as below, except the medium would be incubated at room temperature (22 to 24°C) rather than 35°C, and the incubation time extended to 5 to 7 d as a minimum.

- ☐ Get approval to conduct monitoring in an area where airborne levels of bacteria are likely to be detectable and measurable and perform monitoring accordingly. Most rooms in which people are actively moving will be suitable. Alternatively, conduct a simulated monitoring exercise and obtain a sample spiked with representative bacteria from your instructor for later analysis.
- ☐ At the selected location, disinfect the impactor with alcohol. By holding against a bright light source, examine each stage to check that the holes are all open. If any are blocked, open them by using a nylon or soft plastic bristle or a piece of soft copper or aluminum wire that will easily fit into the hole. Do not use anything that might damage or enlarge a hole.
- ☐ Complete a field monitoring data form, remembering to record name, location or area sampled, date, air temperature, air pressure, relative humidity, analyte (e.g., bacteria), air sampling pump manufacturer and model, pump identification, sampling medium, sample identification, flow rate, start and stop times for sampling, sampling duration, and air volume sampled.

- ☐ Remove covers and position one culture dish with solidified agar in each stage of the disinfected two-stage impactor, assemble the unit, and connect flexible hose.
- ☐ Plug in pump, turn "ON," and let operate for 1 to 2 min prior to connecting flexible hose from the impactor.
- ☐ Place and secure pump in area where sample will be collected. Separate the pump and the impactor adequately so that the impactor is not vibrated by the pump during the collection period.
- ☐ Connect flexible hose to pump.
- ☐ Record the sample identification number and start time. Allow sample to run for at least 10 to 20 min depending on conditions.
- ☐ After an acceptable sampling period has elapsed, turn "OFF" pump, record stop time, and calculate sample time (T [min]).
- ☐ Remove culture dishes from the impactor, cover, and transport samples to laboratory for incubation and analysis.
- ☐ Prepare field blanks by preparing the impactor (disinfecting and loading with two culture dishes) as if another air sample were to be collected. Do not run the pump, but after assembling the impactor, disassemble it and remove the dishes. Identify the dishes as the field blanks and include them in the samples sent or taken to the laboratory. These are incubated and observed as with the sample cultures. They serve as a check of medium sterility and lack of contamination during handling in the field.

3. Post-sampling Calibration of Air Sampling Pump

- ☐ Repeat steps summarized in Method #1, except do not adjust the flow rate of the pump.
- ☐ The post-sampling flow rate (Q_{post}) should be within ±5% of the pre-sampling flow rate (Q_{pre}).
- ☐ Determine the average sampling flow rate (Q_{avg}) by averaging the pre- the post-sampling flow rates. Express average flow rate as liters per minute.

4. Analysis of Sample for Bioaerosol

- ☐ Incubate sample culture dishes upside down at 35°C for 48 h.
- ☐ Place a culture dish on the colony counter and count the total number of colonies.
- ☐ Calculate concentration of both respirable and nonrespirable fractions and express in units of colony forming units per cubic meter.
- ☐ Complete a laboratory analysis data form, remembering to record name, location, date, air temperature, air pressure, pump and impactor manufacturer and model, analytes (e.g., bacteria), incubation conditions, sample/growth medium, sample identification, air volume sampled, and concentration.

UNIT 15 EXAMPLES

Example 15.1

Bioaerosol sampling was performed for bacteria using a TSA medium contained within two petri dishes positioned within a two-stage impactor. The average pre-sampling flow rate was 28.3 l/min and post-sampling flow rate was 28.5 l/min (Q). A sample was collected for 10 min (T [min]). What volume of air was sampled in each location?

Solution 15.1

$$V\ (l) = \left(\frac{28.3\ l/\min + 28.5\ l/\min}{2} \right) \times 10\ \min$$

$$= 284\ l$$

Example 15.2

Following sampling and incubation, the plates were counted. The sample collected in one location had 15 colonies on the agar for the upper (nonrespirable) stage and 19 colonies on the agar for the lower (respirable) stage. Field blank dishes showed no growth. What was the concentration of (a) the nonrespirable fraction, (b) respirable fraction, and (c) total bacteria?

Solution 15.2 (a)

$$\text{Nonrespirable Bacteria } (CFU/m^3) = \frac{15\ CFU}{284\ l} \times \frac{1000\ l}{m^3}$$

$$= 53\ CFU/m^3$$

Solution 15.2 (b)

$$\text{Respirable Bacteria } (CFU/m^3) = \frac{19\ CFU}{284\ l} \times \frac{1000\ l}{m^3}$$

$$= 67\ CFU/m^3$$

Solution 15.2 (c)

$$\text{Total Bacteria } (CFU/m^3) = 53\ CFU/m^3 + 67\ CFU/m^3$$

$$= 120\ CFU/m^3$$

UNIT 16

Evaluation of Airborne Sound Levels: Instantaneous Area Monitoring Using a Sound Level Meter and an Octave Band Analyzer

LEARNING OBJECTIVES

At the completion of Unit 16, including sufficient reading and studying of this and related reference material, learners will be able to correctly:

- Name and identify the common instrument used for conducting instantaneous area monitoring of airborne sound levels.
- Name, identify, and assemble the applicable calibration train that uses an electronic calibrator for calibrating a sound level meter (SLM).
- Calibrate a sound level meter using an electronic calibrator.
- Summarize the principles of monitoring airborne sound levels using an SLM and SLM with octave band analyzer (SLM-OBA).
- Conduct instantaneous monitoring of airborne sound levels using an SLM and an SLM-OBA.
- Record all applicable calibration and sampling data using calibration and field monitoring data forms.

OVERVIEW

Sound waves propagate in a spherical form from the source. If sound measurements are taken at a fixed distance in any direction from a source under free-field, nondirectional conditions (a point source with nothing in the space to impede the sound energy), the sound pressure levels should be the same. In addition, the greater the distance from the source, the lower the sound pressure. Sound pressure changes abide by the inverse square law which states that as the distance increases from the source, there is a corresponding decrease in intensity based on sound level meter readings in decibels. For example, if sound pressure readings are taken x feet from a source, when the distance is doubled to $2x$ ft, a corresponding decrease in sound pressure levels should be observed. Accordingly, if a 100-dB reading on a sound level meter was measured at 2 ft, then a 94-dB reading should be measured at 4 ft. If we are monitoring free-field, nondirectional noise, ideally there should be a 6-dB reduction when the distance doubles. The 6-dB reduction in sound pressure level is the expected pressure level change under free-field, nondirectional conditions.

Since work is rarely performed under ideal conditions, sound pressure changes under actual workplace conditions would be expected to be reduced between 4 and 6 dB in free-field measurement activities as distances double. Sound pressure measures failing to decrease by 4 to 5 dB as the distance doubles from the source indicates that the noise is not free-field, nondirectional noise.

External influences that are augmenting or reflecting/absorbing sound energies in those locations must be determined. Free-field is a noise source that is located in an open area, free of barriers or interferences. An example of an ideal free-field source would be a pole-mounted siren. Free-field noise sources are the exception when monitoring for noise within structures. Ceilings, floors, walls, adjacent equipment, and a wide variety of other sources will act as diffracters, absorbers, or transmitters augmenting or reducing sound pressure levels. As a consequence, barriers will result in sound pressure levels that do not conform with the inverse square law. Sound pressure level contours or noise gradients would then be drawn to determine the effects of structure and contents upon sound pressure changes within the facility.

MONITORING

The SLM and OBA are two generic types of sound monitoring instruments typically used when assessing noise problems in the occupational setting. The SLM is an instrument used to monitor sound pressure levels. The OBA is a special SLM for measuring pressure levels of sound within octave frequency bands. The OBA is actually an electronic filter attached to or incorporated with an SLM. There are a wide variety of SLMs and OBAs commercially available. These instruments are used in the occupational environment to assist in detecting and measuring sound levels and determining whether a noise problem exists. SLMs consist of three major components: (1) a microphone, (2) a meter, and (3) a connection for an OBA (Figure 16.1).

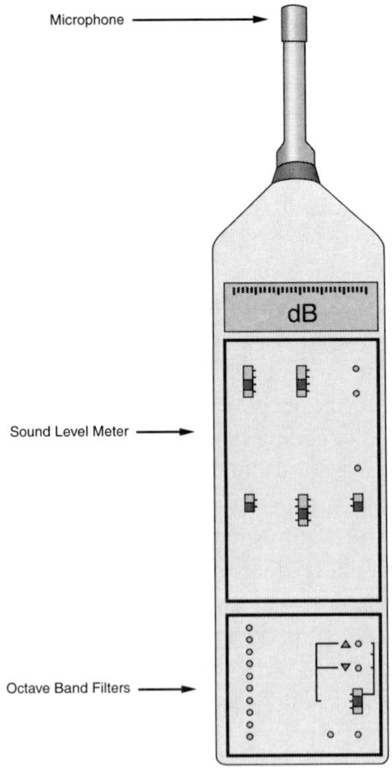

Figure 16.1 SLM with OBA.

While each sound level meter is slightly different, each may have one or more frequency-selective weighting filters or scales in addition to an unweighted filter (also called linear-scale). Three more common scales are labeled A, B, and C, each with characteristics derived from the equal loudness perception of pure tones by humans. The A-scale consists of modified sound pressure levels across the frequency spectrum (20 to 20,000 Hz) which is somewhat close to human hearing response. The A-scale is the mandated scale required for site monitoring activities. The B-scale has limited applications, such as in sound research activities, and is no longer provided in many newly designed SLMs. The C-scale is used for engineering and maintenance activities and particularly is recommended in measuring true impulse (blast-type) sound. Linear-scale is used in octave band analysis and also is mandated by the Occupational Safety and Health Administration (OSHA) regulations for the measurement of impulse sounds.

There are four types of approved SLMs commercially available that differ by their certified accuracy. The Type "0" SLM has an accuracy of ±0 dB and is the most accurate type of SLM available. It is predominantly used for laboratory research applications. The Type 1 SLM is an instrument with an accuracy of ±1 dB. It is a precision unit intended for extremely accurate field and laboratory measurements. The Type 2 SLM is a general purpose SLM with an accuracy of ±2 dB. Type 2 or better SLMs set at A-scale are mandated for use to evaluate the occupational environment for noise. The fourth type of SLM is the Type S or special purpose SLM. It is equivalent to the old Type 3 unit with design tolerances similar to those listed for the Type 1 unit. Type S SLMs differ from Type 1 SLMs because these instruments are not required to have all of the functions of the units previously listed.

While the SLM without OBA can be set at a scale (A, B, C, or linear) that represents a broad range of frequencies, the SLM-OBA can be adjusted to specific frequencies which represent relatively narrow octave band frequencies. The specific "center band frequencies" are 31.5, 63, 125, 250, 500, 1000, 2000, 4000, 8000, and 16,000 Hz. Each center band frequency represents an individual range or band of frequencies. Octave-band frequencies contain progressively wider bandwidth as the center frequency increases. For example, the center band frequency 31.5 Hz is bounded by the lower band-edge frequency of 22.4 Hz and the upper band-edge frequency of 45 Hz; the center band frequency 63 Hz has 45 and 90 Hz as its lower and upper frequency band.

SLMs are calibrated using a known sound level generator called an acoustical calibrator set at a specific sound pressure level (such as 104 dB) and frequency (usually 1000 Hz). The device is simply placed over the microphone of the SLM and the meter is adjusted to the known level (Figure 16.2). It is usually necessary to check calibration of SLM at the beginning and end of noise survey.

When performing the field survey, it is important to correctly hold the SLM. The proper position to hold the SLM is dependent upon the specific instrument to be used or specific application. For example, some instruments require that the unit is held away from the body and at a 90° angle from the sound source, while others may be held at a 70° angle from the source and so forth. When investigating personal exposure, the SLM is usually held at worker's hearing zone.

SOUND GRADIENTS OR CONTOURS

It is commonly advisable to map out the facility where monitoring is conducted, locating all noise sources and indicating the general sound pressure levels surrounding the sources. This map is referred to as a sound level gradient or contour. A sound level gradient is a pictorial representation of the facility with bands of ranked sound pressure levels depicted surrounding the sources.

In the noise gradient evaluation, a scaled diagram of the area is used. Sound pressure level measures are then taken in numerous predetermined grid pattern locations surrounding the noise sources. In addition to the diagram of the perimeter of the room/area, all contents should be included in a noise gradient. Tables, chairs, machines, storage shelves, etc., would be depicted in the sketch. Graph paper is recommended for this task, with each square representing a fixed unit of distance. Using a scale

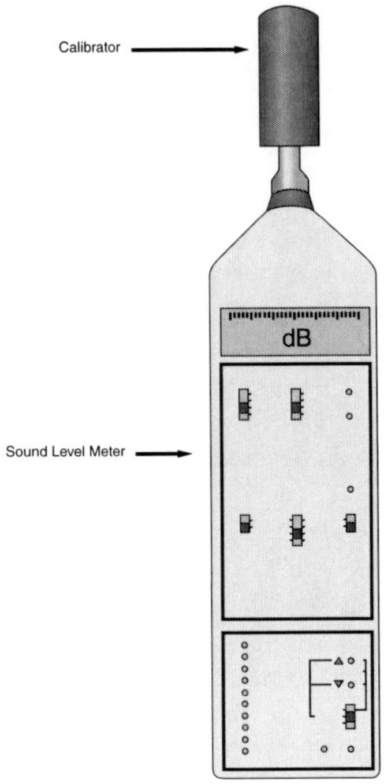

Figure 16.2 Calibration of SLM-OBA by an acoustical calibrator.

of 0.25 in. (one side of a square on standard size graph paper) representing 1 ft, draw to scale the perimeter of the area and all area contents (using labeling codes referenced in the map key).

Employing a tape measure, room contents, and reference points, take several measurements (20 to 40) of sound pressure levels at various locations throughout the area to be evaluated. Once these measurements have been taken and recorded on the area map, connect the measurements with the same sound pressure levels. Rather than noting circular patterns of decreasing noise intensities as would be expected in areas of free-field noise, irregular patterns of sound pressures will emerge.

UNIT 16 EXERCISE

OVERVIEW

The exercise will provide the fundamental concepts for conducting instantaneous area monitoring of sound pressure levels using an SLM-OBA. Place a check mark (✔) in the open box (☐) when you have obtained applicable material and completed the steps for monitoring.

MATERIAL

1. Calibration

☐ Electronic acoustical calibrator
☐ Screwdriver (i.e., 1/8 in.)

2. Monitoring

☐ Type 2 (or better) SLM or SLM-OBA set at A-scale and slow response
☐ Tape measure
☐ Field monitoring data form (Figure 16.3)
☐ 0.25 × 0.25 in. grid graph paper

Field Monitoring Data Form:
Real-Time Monitoring for Sound Levels

Facility Name and Location: ______________________ Date Monitoring Conducted: ____________

Monitoring Conducted By: ______________________

Sound Level Meter (Type/Manufacturer/Model): ______________________

Octave Band Analyzer (Type/Manufacturer/Model): ______________________

Acoustical Calibrator (Type/Manufacturer/Model): ______________________

Pre-Monitoring Calibration: _____ Post-Monitoring Calibration: _____ Meter Response (Fast/Slow): ________

Air Temperature: ____°C Air Pressure: ____mm Hg Relative Humidity: ____%

Identification of Area	Source Distance (ft)	dBA	dBL	31.5 Hz	125 Hz	250 Hz	500 Hz	1 KHz	2 KHz	4 KHz	8 KHz	16 KHz	dBL

Field Notes:

Figure 16.3 Field monitoring data form for real-time area measurements of sound.

METHOD

1. Calibration of the SLM or SLM-OBA

- ☐ Review the manufacturer's instruction manual for proper assembly and operation.
- ☐ Check the batteries for the SLM and the acoustical calibrator.
- ☐ Turn "ON" the SLM and calibrator.
- ☐ Adjust the SLM to select the appropriate decibel range for the acoustical calibrator.
- ☐ Place the calibrator over the microphone of the SLM. If necessary, adjust the meter until it indicates the specific sound pressure level generated by the calibrator. Turn the set screw to adjust the reading.
- ☐ Record data on the field monitoring data form.
- ☐ Remove the calibrator from the SLM and turn "OFF" the acoustical calibrated.
- ☐ The SLM is ready for monitoring.

2. Monitoring Noise Using an SLM or SLM-OBA

- ☐ Get approval to conduct monitoring in an area where airborne levels of sound are likely to be elevated and variable and perform monitoring accordingly. Alternatively, conduct a simulated monitoring exercise using a sound generator as a noise source.
- ☐ Using the SLM and tape measure, measure sound pressure levels at various locations and distances from the source and determine if readings approximate free-field conditions or expectations.
- ☐ Use graph paper and draw a noise gradient or contour map. Include reference points such as a door (e.g., labeled east laboratory entrance) and indicate all noise sources using room contents and structures as reference points. Obtain a floor plan of the facility if possible.
- ☐ Use the SLM-OBA and record measurements at specific center band frequencies in various locations.
- ☐ Complete a field monitoring data form, remembering to record your name, facility and location monitored, date, air temperature, air pressure, relative humidity, and the manufacturer, model, and serial number of each SLM, OBA, and acoustical calibrator.

UNIT 17

Evaluation of Airborne Sound Levels: Integrated Personal Monitoring Using an Audio Dosimeter

LEARNING OBJECTIVES

At the completion of Unit 17, including sufficient reading and studying of this and related reference material, learners will be able to correctly:

- Name and identify the common instrument used for conducting instantaneous personal monitoring of airborne sound levels.
- Name, identify, and assemble the applicable calibration train that uses an electronic calibrator for calibrating an audio dosimeter.
- Calibrate an audio dosimeter using an electronic calibrator.
- Summarize the principles of monitoring airborne sound levels using an audio dosimeter.
- Conduct instantaneous personal monitoring of airborne sound levels using an audio dosimeter.
- Conduct all applicable calculations for allowable time, dose, and sound pressure level.
- Record all applicable calibration and sampling data using field monitoring data forms.

OVERVIEW

In addition to instantaneous or real-time area monitoring using sound level meters (SLMs), integrated or continuous personal monitoring using audio dosimeters is frequently conducted to detect and measure sound levels when evaluating the occupational environment. Following monitoring, calculations can be performed to determine whether levels are within standard or recommended criterion levels. The Occupational Safety and Health Administration (OSHA) Noise Standard (29 CFR 1910.95) has established requirements relative to monitoring occupational environments for determination of excessive exposures to elevated sound pressure levels. A criterion level is the continuous A-weighted sound level which constitutes 100% of an allowable exposure. For example, as per the OSHA Noise Standard, the criterion level is 90 dBA for an 8-h workday. If noise exposures greater than 90 dBA are based upon a 5-dB "doubling exchange rate," then workers are permitted to be exposed to 95 dBA for 4 h, 100 dBA for 2 h, 105 dBA for 1 h, and so on. Allowable time (T [h]) can be calculated for exposure to a given sound pressure level (L_a) based on a criterion level of 90 dBA and a doubling exchange rate of 5.

AF uses 3-dB

$$T\,(\mathrm{h}) = \frac{8\ \mathrm{h}}{2^{(L_a - 90)/5}} \tag{17.1}$$

To calculate noise exposure in locations or job classifications where sound pressure levels vary, the total sound pressure exposure or dose must be determined. Dose (D) is calculated based on the actual exposure time at a given sound pressure level (C) divided by the allowable exposure time at that level.

$$D\% = \left(\frac{C_1}{T_1} + \frac{C_2}{T_2} + \cdots + \frac{C_n}{T_n}\right) \times 100 \qquad (17.2)$$

If the total dose exceeds 100% or unity, then individuals exposed to the various sound sources at the measured levels would be excessively exposed.

To determine the equivalent sound pressure level (L_{eq}) in dBA for an 8-h workday based upon dose (D) values, Equation 17.3 is used.

$$L_{eq} = 90 + 16.61 \log\left(\frac{D\%}{100}\right) \qquad (17.3)$$

To determine the equivalent sound pressure level (L_{eq}) in dBA for dose (D) exposure time (T [h]) is not equal to an 8-h shift, Equation 17.4 is used.

$$L_{eq} = 90 + 16.61 \log\frac{D\%}{12.5\ T} \qquad (17.4)$$

MONITORING

Audio dosimeters are special SLMs that integrate sound pressure levels over time and are used to measure personal noise exposures. Audio dosimeters are composed of a microphone attached to a meter that is similar to a sound level meter (Figure 17.1). The detected sound levels,

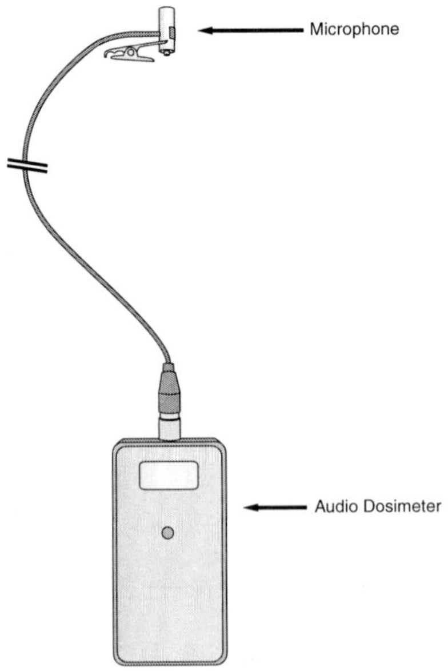

Figure 17.1 Audio dosimeter.

however, are integrated into memory, and measurements are indicated via direct readout. Computerized readout devices are also available. The audio dosimeter microphone is typically clipped to the worker's clavicle (collar) or trapezius (upper shoulder) region, provided it is within 12 to 18 in. of the ear (i.e., hearing zone). Ideally, the microphone of the device should be connected to the worker's ear, but this is not always practical nor desired. The meter is either placed in the shirt pocket or attached to the belt. Audio dosimeters integrate the A-scale sound pressures that the worker is exposed to over an entire shift (or monitoring period), and total exposure or dose is then determined.

By calculating the integrated dose obtained with the audio dosimeter and inserting this value into one of the equivalent sound pressure level formulas presented earlier, it is possible to determine the level of exposure. Modern audio dosimeters also can be downloaded to a computer and generate printouts showing a direct readout of a wide variety of measures obtained throughout the monitoring period. Most of these same instruments also have a direct readout display on each unit.

UNIT 17 EXERCISE

OVERVIEW

The exercise will provide the fundamental concepts for conducting integrated personal monitoring of sound levels using an audio dosimeter. Place a check mark (✔) in the open box (☐) when you have obtained applicable material and completed the steps for sampling.

MATERIAL

1. Calibration

- ☐ Electronic acoustical calibrator
- ☐ Screwdriver (i.e., 1/8 in.)

2. Monitoring

- ☐ Audio dosimeter
- ☐ Field monitoring data form (Figure 17.2)

METHOD

1. Calibration of an Audio Dosimeter

- ☐ Like the sound level meter, the correct operation of an audio dosimeter is dependent upon the specific unit in question. As a consequence, the manufacturer's instruction manual must be reviewed for proper calibration and operation.
- ☐ Check the batteries for the electronic calibrator and the audio dosimeter.
- ☐ Turn "ON" the calibrator and audio dosimeter.
- ☐ Insert the microphone into the calibrator until it indicates the specific intensity generated by the calibrator. Turn the set screw to adjust the setting.
- ☐ Remove the audio dosimeter microphone from the calibrator and turn "OFF" both devices.

Field Monitoring Data Form:
Integrated Monitoring for Sound Levels

Facility Name and Location: ______________________ **Date Monitoring Conducted:** __________

Monitoring Conducted By: ______________________

Audio Dosimeters and Calibrator (Manufacturer/Model): ______________________ **Cut-Off:** ________

Air Temperature: ____°C **Air Pressure:** ____mm Hg **Relative Humidity:** ____%

Identification of Personnel or Area	Audio Dosimeter No.	Pre-Monitoring Calibration	Post-Monitoring Calibration	Sample Start-Time	Sample Stop-Time	Total Sample Time (T)	Dose (%)	Sound Pressure Level (L_{eq}) (dBA)

Field Notes:

Figure 17.2 Field monitoring data form for integrated measurements of sound.

2. Monitoring Using an Audio Dosimeter

- ☐ Identify a noisy area and connect the audio dosimeter to an individual. For the purpose of this exercise, either have the individual wear the audio dosimeter in a noisy environment for at least 2 h or place the microphone near an identified noise source for an appropriate time period to spike the device for demonstration purposes.
- ☐ Read the measurement and record data.
- ☐ Complete a field monitoring data form, remembering to record your name, facility and location monitored, date, air temperature, air pressure, relative humidity, and audio dosimeter and calibrator manufacturer and model.

UNIT 17 EXAMPLES

Example 17.1

What is the maximum allowable time for exposure to 96 dBA?

Solution 17.1

$$T\ (\text{h}) = \frac{8\ h}{2^{(L_a - 90)/5}}$$

$$= \frac{8\ h}{2^{(96-90)/5}}$$

$$= 3.48\ h$$

Example 17.2

What is the percent dose (D%) for a 3-h exposure to 90 dBA, a 2-h exposure to 95 dBA, and a 3-h to 100 dBA?

Solution 17.2

$$D\% = \left(\frac{C_1}{T_1} + \frac{C_2}{T_2} + \frac{C_3}{T_3}\right) \times 100$$

$$= \left(\frac{3}{8} + \frac{2}{4} + \frac{3}{2}\right) \times 100$$

$$= 237.5\%$$

Example 17.3

What is the equivalent sound pressure level (L_{eq}) for a dose of 237.5%?

Solution 17.3

$$L_{eq} = 90 + 16.61 \ \log\left(\frac{D\%}{100}\right)$$

$$= 90 + 16.61 \ \log\left(\frac{237.5\%}{100}\right)$$

$$= 96 \text{ dBA for 8 h}$$

UNIT 18

Evaluation of Personal Hearing Thresholds: Instantaneous Personal Monitoring Using an Audiometer

LEARNING OBJECTIVES

At the completion of Unit 18, including sufficient reading and studying of this and related reference material, learners will be able to correctly:

- Name and identify the instrument used for conducting instantaneous monitoring of human hearing thresholds.
- Summarize the principles of monitoring human hearing thresholds.
- Conduct instantaneous monitoring of human hearing thresholds using an audiometer.
- Perform applicable calculations relative to standard threshold shifts.
- Record all applicable monitoring data using a field monitoring data form.

OVERVIEW

As already defined in previous units, noise may be simply described as unwanted sound. Sound is a form of vibration which may be conducted through solids, liquids, or gases. It is an energy form in the air. The three important properties of noise that may result in hearing loss are intensity or pressure, frequency, and duration. The louder the noise, the higher the intensity. Higher frequencies (commonly described as high pitches) tend to be more damaging than lower pitched noise. The longer the exposure, the greater the likelihood of prolonged and sustained damage. According to The Occupational Safety and Health Administration (OSHA) Noise Standard, audiometry shall be provided by the employer when employees are exposed to noise ≥85 dBA as an 8-h time-weighted average.

Audiometry is the measurement of hearing thresholds of humans when exposed to known pure tone sounds at specific frequencies. Audiometric monitoring is not a routine responsibility of industrial hygienists. In all likelihood, the occupational health and safety professional will not have the credentials to conduct audiometric monitoring. Audiometry should be conducted by a certified audiologist. Alternatively, testing can be conducted by an occupational hearing conservationist certified by the Council for Accreditation in Occupational Hearing Conservation (CAOHC) or equivalent and supervised by a physician or audiologist[1]. Although typically not responsible for actually conducting the testing, many industrial hygienists are expected to be familiar with major aspects of audiometric testing. They typically are responsible for or involved with administration of hearing conservation programs. To assure that valid and reliable data are collected, only certified personnel should be responsible for conducting audiometric tests. However, occupational health professionals should be knowledgeable about audiometric monitoring protocols, as well as interpretation of audiometric data.

MONITORING

Audiometers are instruments that generate a known intensity of sound at a known frequency. The sound is conducted from the audiometer through headphones worn by the individual being evaluated (Figure 18.1). The audiometers can be controlled to generate sound to each ear separately. Several types of audiometers are commonly used to conduct audiometric monitoring. The audiometric test consists of air-conduction, pure-tone, and hearing threshold measures. The pure-tone audiometer is a manually operated audiometer that is frequently used in the occupational setting. The frequency, intensity, and tone presentation are controlled by the audiologist or certified technician. While there may be other features on these units, alternative features are primarily used in diagnostic testing and are not typically intended for use when evaluating occupational noise-induced hearing loss.

Ideally, sound controlled or audiometric booths should be used when conducting audiometric evaluations. It is permissible, however, to test in open rooms, but the room location and its proximity to loud noise are important factors to consider. Also, individuals being tested must not be exposed

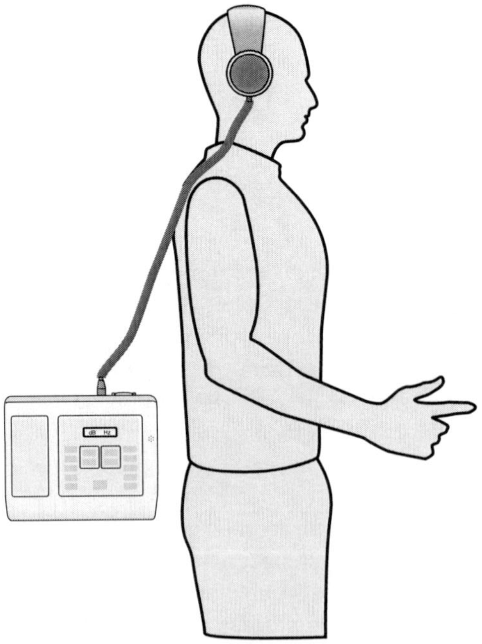

Figure 18.1 Audiometer with headphones connected to subject.

[1] U.S. Department of Health and Human Services, Criteria for a Recommended Standard: Occupational Noise Exposure, NIOSH Publication NO. 98-126 (revised criteria 1998), National Institute for Occupational Safety and Health, Cincinnati, OH, June 1998.

to sound levels ≥85 dBA for a minimum of 12 h prior to obtaining a baseline audiogram. Baseline audiometric evaluation of workers is a common component of pre- and initial employment, occupational medical surveillance programs. In addition, exit audiometric testing may be conducted when a worker is either transferred to a different job or employment is terminated.

The output from audiometric testing is called an audiogram. Audiograms are standard charts or graphs representing the minimum intensity or threshold level at which various sounds produced by an audiometer are just barely heard by the individual tested. These detection points are referred to as hearing thresholds and are measured at a range of individual frequencies.

Horizontal rows across the top of the audiogram are the frequencies, labeled in hertz (Hz). For occupational monitoring purposes, frequencies evaluated include 250, 500, 1000, 2000, 3000, 4000, 6000, and 8000 Hz. Vertically, the numbers along the left column of an audiogram represent loudness or intensity, measured in decibels (dB). Audiograms should be laid out graphically to ensure that 20 dB column intervals and row intervals are equal, forming a square (Figure 18.2). Threshold values are recorded for each frequency by marking the appropriate location with an "x" (and if color coded using blue), representing the results for the left ear, and an "o" (and if color coded using red), representing the results for the right ear.

An audiogram for an individual is compared to audiograms recorded previously (e.g., pre-employment baseline audiogram). Taking values at 2000, 3000, and 4000 HZ, for example, a standard threshold shift (STS) can be calculated based on changes in hearing thresholds at each of the three frequencies. STS is calculated for each ear. A calculated STS of >10 dB for either ear may be considered a significant indication of hearing loss.

$$\text{STS} = \frac{\left(\Delta dB_{2000\ Hz} + \Delta dB_{3000\ Hz} + \Delta dB_{4000\ Hz}\right)}{3} \tag{18.1}$$

Audiogram Form

250 Hz	500 Hz	1000 Hz	2000 Hz	4000 Hz	8000 Hz	
						0 dB
						10 dB
						20 dB
						30 dB
						40 dB
						50 dB
						60 dB
						70 dB
						80 dB
						90 dB
						100 dB

Figure 18.2 Audiogram form.

Workers determined to have an STS need to be protected from further noise-induced impact. Protective measures include reassignment of the worker to a quieter area, refitting of hearing protectors (e.g., earplugs, earmuffs), and additional training on hearing loss and prevention.

UNIT 18 EXERCISE

OVERVIEW

The exercise will provide the fundamental concepts for personal monitoring of hearing by conducting air-conduction pure-tone audiometric testing using an audiometer. Place a check mark (✔) in the open box (☐) when you have obtained applicable material and completed the monitoring steps.

MATERIAL

1. Calibration

☐ Audiometers are periodically factory calibrated, although field calibration can be conducted.

2. Monitoring

☐ Audiometer
☐ Field monitoring data form (Figure 18.3)

METHOD

1. Calibration of an Audiometer

☐ Conducted by manufacturer

2. Monitoring Using an Audiometer

☐ Read the manufacturer's instruction for assembly and operation of the audiometer.
☐ Summarize the testing procedure to the subject.
☐ Fit the red phone to the right ear and the blue phone to the left ear.
☐ Begin by testing the better ear, if known. If unknown, start by testing the right ear.
☐ Present a 1000 Hz tone at 40 to 50 dB above the subject's presumed threshold for 1 to 3 sec. If the individual does not respond, repeat and increase the intensity until you get a response.
☐ Reduce the intensity successively in 10-dB steps and vary the interval between presentations of tones to avoid anticipatory responses.
☐ After reaching an intensity level at which no responses are made, increase the level in 5-dB steps until the subject responds once more.
☐ Decrease the intensity in 10-dB steps and increase in 5-dB steps until the threshold is evident (the individual responds to 50% or 2 of 4 trials at the intensity of concern).
☐ Repeat the previous steps for 2000, 3000, 4000, 8000, 500, and 250 Hz in that order.
☐ Check the 1000-Hz threshold again before switching to the other ear, which is tested in the exact same way.
☐ Record data on the field monitoring data form and prepare an audiogram.
☐ Complete the field monitoring data sheet, remembering to record name of technician, name of subject, date, location, and audiometer manufacturer and model.

Audiometric Monitoring Data Form:
Real-Time Monitoring for Hearing Thresholds

Facility Name and Location: ______________________ **Date Monitoring Conducted:** ______________

Monitoring Conducted By: ______________________

Audiometer (Type/Manufacturer/Model): ______________________________________

Monitor Calibration: ______________________

Identification of Personnel	Age (yr)	Hearing Threshold (dB) Per Ear (R and L) at Specific Frequency (Hz)															
		250 Hz		500 Hz		1000 Hz		2000 Hz		3000 Hz		4000 Hz		6000 Hz		8000 Hz	
		R	L	R	L	R	L	R	L	R	L	R	L	R	L	R	L

Comments:

Figure 18.3 Field monitoring data form for an audiometric evaluation.

UNIT 18 EXAMPLE

Example 18.1

Data from audiometric baseline and follow-up evaluations of a worker are shown below. What was the STS and was it significant?

Frequency (Hz)	Baseline Audiogram (dB)	Follow-Up Audiogram (dB)	Change (dB)
500	5	5	0
1000	5	5	0
2000	0	10	+10
3000	5	15	+10
4000	10	30	+20
6000	10	15	+5

Solution 18.1

$$STS = \frac{(^{+}10\ \text{dB}_{\Delta 2000\,Hz} + ^{+}10\ \text{dB}_{\Delta 3000\,Hz} + ^{+}20\ \text{dB}_{\Delta 4000\,Hz})}{3}$$

$$= 13.3\ \text{dB}$$

Since 13.3 dB > 10 dB, the STS is significant.

UNIT 19

Evaluation of Heat Stress: Instantaneous Area Monitoring Using a Wet-Bulb Globe Temperature Assembly and Meter

LEARNING OBJECTIVES

At the completion of Unit 19, including sufficient reading and studying of this and related reference material, learners will be able to correctly:

- Name, identify, and assemble the common instruments used for conducting instantaneous monitoring of the parameters for determining the wet-bulb globe temperature (WBGT) index.
- Summarize the principles of monitoring for heat stress indices.
- Conduct instantaneous measurements of dry-bulb (DB) temperature, globe temperature (GT), and natural wet-bulb (NWB).
- Perform applicable calculations and conversions relative to the WBGT index indoors ($WBGT_{in}$) and outdoors ($WBGT_{out}$).
- Record all applicable sampling data using field monitoring data forms.

OVERVIEW

Several devices and procedures are available for evaluating the occupational environment for conditions that may contribute to heat stress. The most common method involves determination of the WBGT index. The method involves the collection of two to three major measurements that reflect the contribution of environmental factors to heat stress.

The measurements include the DB temperature, NWB temperature, and the GT. The DB temperature measures the ambient temperature. The NWB temperature records the ambient temperature, but reflects the influence of humidity and air movement. High humidity and low air movement result in an elevated WB temperature. The GT reflects the contribution of radiant heat. An integration of all three measurements yields a WBGT index. For indoor and outdoor settings where there is no solar load, only the NWB and GT measurements are needed.

$$WBGT_{in} = (0.7 \times NWB) + (0.3 \times GT) \tag{19.1}$$

For outdoors in sunshine, the air temperature based on the DB temperature must also be measured.

$$WBGT_{out} = (0.7 \times NWB) + (0.2 \times GT) + (0.1 \times DB) \tag{19.2}$$

Although the WBGT is easy and simple to use, it does not in itself include a factor for the work rate or the clothing of an individual. It is not possible to determine an allowable exposure time directly from the WBGT. Guidelines have been developed that eliminate these difficulties. The workload for an individual is determined by estimating the worker's metabolic rate. A chart has been developed to aid in this process. There is also a table to make appropriate modification to WBGT to account for the clothing effects. An occupational exposure limit has been presented based on the amount of time a worker is involved in continuous work, the level of work that the worker is performing, and the types of clothing worn.

MONITORING

Although measurement of the ambient air temperature by itself is insufficient to determine heat stress, it is an integral part of heat stress determinations. There are various methods to assess ambient air temperature. The most common method of measurement is the DB thermometer. A DB thermometer consists of a hollow glass tube with a bottom reservoir of mercury. The range of the thermometer should be –5 to +50°C and accurate to ±0.5°C. Temperature beyond the range of a DB thermometer may break the thermometer. When the measurements are taken, the DB thermometer must be shielded from radiant heat sources so that only the temperature of the ambient air will be detected. Thermoelectric thermometers or thermocouples are also commonly used to measure the ambient air temperature.

The black globe thermometer, or Vernon globe, is the standard method for measuring radiant heat. A black globe thermometer consists of a 6-in. diameter hollow copper sphere that is painted matte black. A thermometer is inserted so that its bulb is centered inside the globe. The range of the thermometer should be –5 to +100°C and accurate to ±0.5°C. The black globe absorbs radiant heat increasing the temperature of air within the globe proportional to the amount of heat energy absorbed. The thermometer inside the globe is allowed to reach equilibrium. The temperature calculated from GT is termed the mean radiant temperature, which is indicative of the average temperature of surrounding environment.

The NWB temperature is measured by a thermometer with its bulb (reservoir of mercury) wrapped over with a wet wick. The wick is made of a highly absorbent woven cotton. The covered thermometer bulb is partially immersed in distilled water to keep the wick wet. As the wick evaporates water, a certain amount of heat energy is dissipated through evaporative cooling. The bulb is then cooled by the heat absorbed by water during evaporation of the water, and equilibrium is reached between the evaporation rate and the water vapor pressure in the air. The temperature is indicated on the thermometer. At full saturation, the WB thermometer temperature will equal the DB temperature since no evaporative cooling will be experienced. The NWB thermometer is only exposed to natural (not forced) air movement and is not shielded from radiation.

As the temperature of air rises, it will hold more water vapor. When the air at a given temperature holds all the water vapor that it can, it is said to be saturated. The water in the air exerts a vapor pressure which is measured in millimeters of mercury or pascal. The relative humidity is the percentage of water vapor in the air relative to the amount the air would hold when saturated at the same ambient temperature. Relative humidity in the air is usually measured using a DB thermometer side-by-side with a WB thermometer. A sling psychrometer (Figure 19.1) combines both thermometers. The psychrometer is whirled, thus passing air over the bulbs of the thermometers. The heat loss by evaporation keeps the WB thermometer in lower temperature relative to that obtained by the DB thermometer. Relative humidity is then determined using the psychrometric chart. This chart provides relationships between relative humidity, DB temperature, WB temperature, ambient water vapor pressure, and dew point temperature. The thermometers and accessories needed to measure WB temperature, GT, and DB temperature can be assembled and positioned in a ring stand. More commonly today, commercially available electronic WBGT meters are used (Figure 19.2). The electronic devices can integrate the temperature measurements and provide a direct readout for individual temperatures and the WBGT.

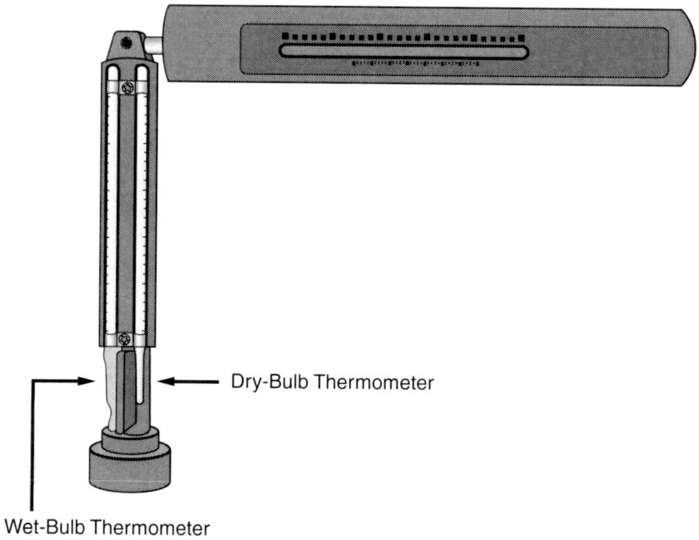

Figure 19.1 Sling psychrometer.

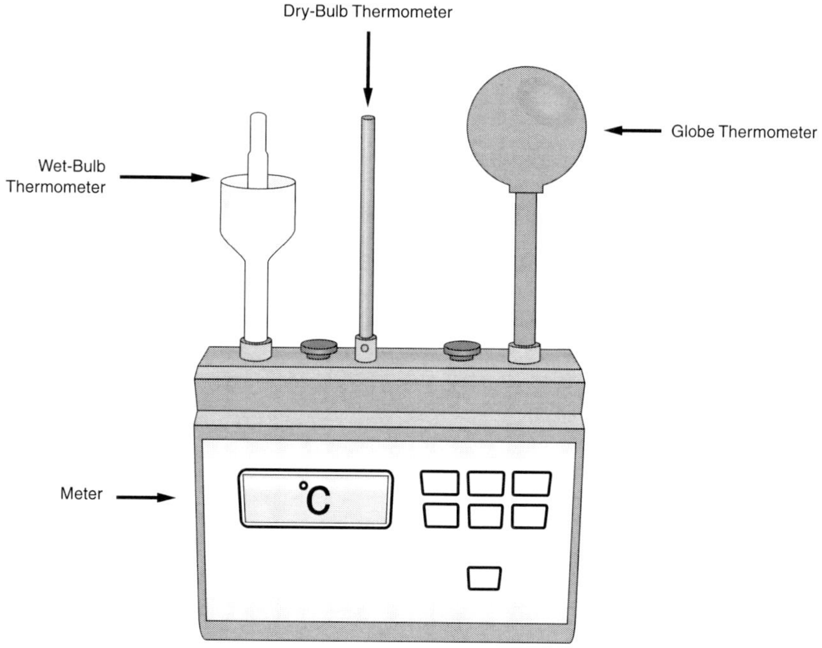

Figure 19.2 Electronic meter for measuring WBGT.

UNIT 19 EXERCISE

OVERVIEW

The exercise will provide the fundamental concepts for conducting instantaneous area monitoring of heat stress parameters using a manual WBGT assembly or an electronic WBGT meter. Place a check mark (✔) in the open box (☐) when you have obtained applicable material and completed the steps for monitoring.

MATERIAL

1. Monitoring WBGT

- ☐ DB thermometer
- ☐ WB thermometer
- ☐ NWB thermometer
- ☐ GT thermometer
- ☐ Standard laboratory ring stand with clamps
- ☐ In addition or as an alternative to the above, a commercially available, battery-operated, electronic WBGT meter can be used to measure the three temperatures
- ☐ Sling psychrometer
- ☐ Field monitoring data form (Figure 19.3)

METHOD

1. Monitoring WBGT

- ☐ If the manual method is used, position the three thermometers vertically on the ring stand using a clamp for each, assuring that the WB wick is immersed in deionized water. For the electronic WBGT meter, assemble the DB, NWB, and GT probes, fill the NWB reservoir with deionized water, and turn "ON."
- ☐ Position and secure the WBGT assembly in an outdoor area and then an indoor area for 20 min in each location.
- ☐ For each area, read and record the temperatures indicated on each thermometer, remembering to use consistent units of Fahrenheit or Celsius. Calculate the $WBGT_{in}$ and $WBGT_{out}$ for each area. If an electronic WBGT meter is used, read and record the individual temperatures and the WBGT automatically from the direct readout display.
- ☐ For each area, operate, read and record DB and WB, and using the psychrometric chart determine the relative humidity.
- ☐ Complete the field monitoring data form, remembering to record name, location, air temperature, atmospheric pressure, relative humidity, and instrument manufacturer, type, and model.

UNIT 19 EXAMPLES

Example 19.1

What is the indoor WBGT index if the measured DB temperature was 78°F, the natural WB temperature was 58°F, and the globe temperature was 80°F?

Field Monitoring Data Form:
Real-Time Monitoring for WBGT Indices

Facility Name and Location: ______________________ **Date Monitoring Conducted:** ____________

Monitoring Conducted By: ______________________

Monitoring Instrument (Type/Manufacturer/Model): ______________________

Ambient Air Temperature: ____°C **Air Pressure:** ____mm Hg **Relative Humidity:** ____%

Identification of Area	Sample Start-Time	Sample Stop-Time	Total Sample Time (T)	Dry Bulb Temperature (°C)	Wet Bulb Temperature (°C)	Black Globe Temperature (°C)	WBGT (°C)

Field Notes:

Figure 19.3 Field monitoring data form for WBGT measurements.

Solution 19.1

$$\begin{aligned} \text{WBGT}_{\text{in}} &= (0.7 \times NWB) + (0.3 \times GT) \\ &= (0.7 \times 58°F) + (0.3 \times 80°F) \\ &= 64.6°F \end{aligned}$$

Example 19.2

What is the outdoor WBGT index for the temperature indicated in Example 19.1?

Solution 19.2

$$\begin{aligned} \text{WBGT}_{\text{out}} &= (0.7 \times NWB) + (0.2 \times GT) + (0.1 \times DB) \\ &= (0.7 \times 58°F) + (0.2 \times 80°F) + (0.1 \times 78°F) \\ &= 64.4°F \end{aligned}$$

UNIT 20

Evaluation of Illumination: Instantaneous Area Monitoring Using a Light Meter

LEARNING OBJECTIVES

At the completion of Unit 20, including sufficient reading and studying of this and related reference material, learners will be able to correctly:

- Name, identify, and assemble the common instruments used for conducting instantaneous monitoring of illumination levels.
- Summarize the principles of monitoring for illumination.
- Conduct instantaneous measurement of illumination.
- Record all applicable sampling data using a field monitoring data form.

OVERVIEW

Visible light is part of the electromagnetic waves spectrum and nonionizing radiation. Light is transmitted linearly through air. The direction of light can be changed, however, depending on the interaction with various media, and incident light will scatter or reflect from some surfaces. When light passes through different transparent media, it will bend or refract. In addition, when light passes through and exits a transparent medium, it will bend and diffuse.

Lighting is important in the occupational environment for reasons such as worker comfort and safety. Several factors influence the quality and quantity of lighting. Quality of light refers to factors such as contrast, brightness, glare, and color. Illumination is directly related to the quantity of the light energy falling on a given area. Illumination is evaluated in the occupational environment to determine if lighting quality and quantity are appropriate for specific and general tasks.

Visible light is generated from both natural and anthropogenic sources. Luminous flux refers to the total energy from a source of light that is emitted and can influence illumination and sight. The luminous flux is measured in standard units called lumen. One lumen is the radiant power emitted from a standard source. Illumination is the density of the luminous flux on a given area of a surface. Thus, illumination equals lumens per area. Illumination is measured either in old units of foot-candle or in metric system units of lux. One foot-candle equals approximately 10 luxes.

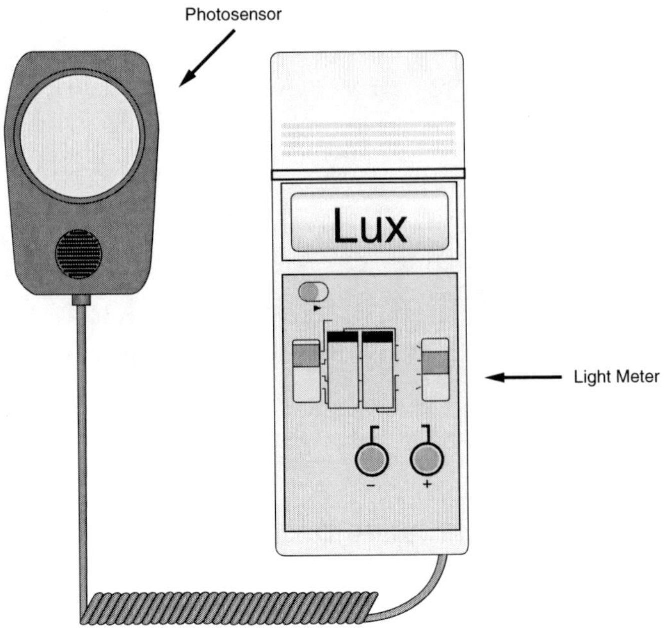

Figure 20.1 Light meter.

MONITORING

An electronic light meter is used to measure illumination. The device consists of a relatively flat photosensitive sensor with a color attenuator or filter connected to a detector (Figure 20.1). Light energy causes emission of electrons and related generation of increased current. The current generated is directly related to the intensity of light energy. Measurement of illumination, in foot-candles or lux, is indicated in a direct readout display. The light meter is put on the surface of task with the detector facing upward.

Illumination levels are measured and recorded at various points on a given surface where the task is performed. Illumination is monitored at the specific work surfaces and also in the general areas of a facility.

UNIT 20 EXERCISE

OVERVIEW

The exercise will provide the fundamental concepts for conducting instantaneous monitoring of illumination generated from a source of visible light using an applicable electronic light meter. Place a check mark (✔) in the open box (☐) when you have obtained applicable material and completed the steps for monitoring.

MATERIAL

1. Monitoring for Illumination

- ☐ Light meter
- ☐ Tape measure
- ☐ Natural and artificial light sources
- ☐ Field monitoring data form (Figure 20.2)

Field Monitoring Data Form:
Real-Time Monitoring for Illumination
Using An Illumination Meter

Facility Name and Location: ______________________________

Monitoring Conducted By: ______________________________

Date Monitoring Conducted: ______________

Monitoring Instrument (Type/Manufacturer/Model): ______________

Natural Light Sources: ______________________________

Artificial Light Sources: ______________________________

Identification of Area	Illumination (footcandles)

Field Notes:

Figure 20.2 Field monitoring data form for illumination measurements.

METHOD

1. Monitoring for Illumination

- ☐ Refer to the light meter manufacturer's instructions for assembly and operation of the instrument. Note that the meter is factory calibrated and must be returned to the manufacturer or the equivalent for periodic adjustments and recalibration.
- ☐ Measure the dimensions of the illuminated area and distances from various natural and artificial sources of light.
- ☐ Turn "ON" the light meter.
- ☐ Hold the sensor away from the body and on the working surfaces or floor and record the illumination levels measurement.
- ☐ Complete a field monitoring data form, remembering to record name, location, date, light sources, and light meter manufacturer and model.

UNIT 21

Evaluation of Airborne Microwave Radiation: Instantaneous Monitoring Using a Microwave Meter

LEARNING OBJECTIVES

At the completion of Unit 21, including sufficient reading and studying of this and related reference material, learners will be able to correctly:

- Name, identify, and assemble the common instrument used for conducting instantaneous monitoring of microwave radiation.
- Summarize the principles of monitoring for microwave radiation.
- Conduct instantaneous measurement of microwave radiation.
- Record all applicable sampling data using a field monitoring data form.

OVERVIEW

Microwaves are electromagnetic nonionizing radiation with frequencies ranging from 300 MHz to 300 GHz corresponding to wavelengths from 1 m to 1 mm, respectively. The microwave energy is generated by the acceleration of electrons in oscillatory circuits. Depending on the frequency (or wavelength), microwaves are transmitted for long distances through air. A major characteristic of microwave energy is that it can be readily absorbed by numerous materials, including human tissue, resulting in the generation of elevated temperatures within a given volume. Relative to human tissue, microwaves are more penetrating in relation to areas of higher water content, such as the eyes, skin, muscles, and visceral organs. In addition, higher frequency and shorter wavelength can increase the depth of absorption.

Common uses and sources of microwave radiation include dielectric heaters, radio frequency sealers, radio and television broadcast stations, induction furnaces, plasma processing, electrosurgical units, communication systems (i.e., radar), and ovens. Microwave ovens are a common use and source of microwaves in both occupational and nonoccupational environments. Poorly aligned doors, accumulated food residues, and faulty door gaskets, hinges, and latches can result in the release of microwave radiation from inside to outside the microwave oven. Leaking microwave ovens can generate significant and dangerous amounts of microwave radiation into the external air. Accordingly, the ovens should be monitored periodically to determine if microwave radiation is detected and measured outside of a closed and operating unit.

The power density of microwave radiation is actually measured and expressed in units of milliwatts per square centimeter (mW/cm^2). Microwave oven performance standards specify less than 1 mW/cm^2 at any point 5 cm from the oven before purchase and less than 5 mW/cm^2 at any point 5 cm from the oven after purchase.

MONITORING

Several types of microwave detectors are available, including thermal devices that respond to increased temperatures from absorbed radiation by reflecting a change in resistance or voltage. Electronic microwave meters that convert microwave frequency into direct current are commonly used for instantaneous area monitoring of nonionizing radiation from microwave ovens. The meters consist of an antenna or probe that is connected to a detector and amplifier (Figure 21.1). A spacer may be present on the probe so that a uniform distance is maintained between the source and the antenna. When evaluating a microwave oven, a 5-cm distance is required between the oven door and the probe during monitoring.

Prior to initiating the monitoring procedure, a thorough inspection of the microwave oven should be conducted. This inspection should include a check of the electrical wiring and plug, door alignment, door seal condition and cleanliness, and door hinge and latch condition. Following inspection, materials used for the standard loading of the unit must be placed in the microwave oven. Operation of ovens with no load may cause damage to the unit and must be avoided. Accordingly, a standard load typically consists of a 500-ml nonmetal (i.e., plastic) beaker filled with approximately 275 ml of tap water. Monitoring is performed while the oven is operating by keeping the probe 5 cm from the surface of the unit and scanning the door area (Figure 21.2). Following the evaluation, a thermometer is immersed in the beaker of water to verify an increase in the temperature of the water due to microwave energy.

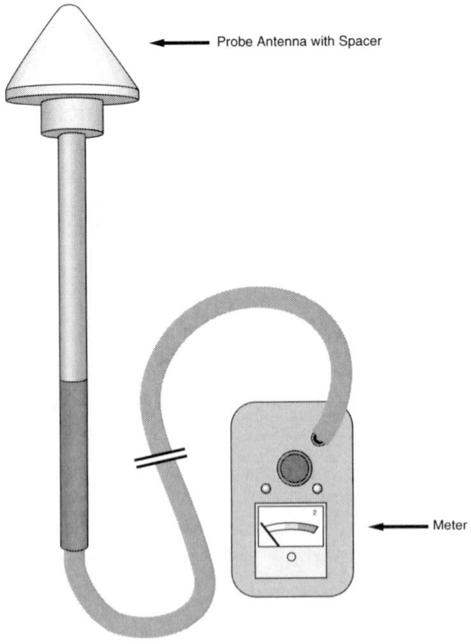

Figure 21.1 Microwave radiation survey meter.

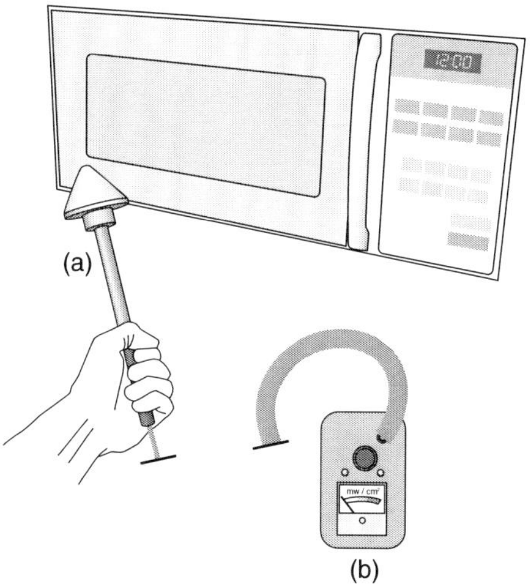

Figure 21.2 Microwave survey meter assembly consisting of (a) a probe and (b) a readout for measuring a microwave oven.

UNIT 21 EXERCISE

OVERVIEW

The exercise will provide the fundamental concepts for conducting instantaneous monitoring of microwave radiation generated from a microwave oven using an applicable electronic survey meter. Place a check mark (✔) in the open box (☐) when you have obtained applicable material and completed the steps for the monitoring method.

MATERIAL

1. Monitoring a Microwave Oven for Leaking Microwave Radiation

☐ Microwave survey meter
☐ 500-ml plastic or glass microwave-safe beaker
☐ 275-ml tap water
☐ Thermometer (20 to 120°C)
☐ Microwave oven
☐ Tape measure or ruler
☐ Field monitoring data form (Figure 21.3)

METHOD

1. Monitoring a Microwave Oven for Leaking Microwave Radiation

☐ Refer to the survey meter manufacturer's instructions for assembly and operation of the instrument. Note that the meter is factory calibrated and must be returned to the manufacturer or the equivalent for periodic adjustments and recalibration.

Field Monitoring Data Form:
Real-Time Monitoring for Microwave Radiation
from an Oven Using A Microwave Oven Survey Meter

Facility Name and Location: ______________________________

Monitoring Conducted By: ______________________________

Date Monitoring Conducted: ______________

Monitoring Instrument (Type/Manufacturer/Model Number): ______________

Microwave Oven (Type/Manufacturer/Model): ______________

Water Volume (Load): ____ml

Water Pre-Temperature: _____ °C **Water Post-Temperature** _____ °C

Sample Point	Level (mW/cm^2)	Microwave Oven Diagram and Sample Points
1		
2		
3		
4		
5		
6		
7		
8		
9		
10		
11		
12		

Field Notes:

Figure 21.3 Field monitoring data form for microwave radiation measurements.

- ☐ Assemble the survey meter by attaching the probe (antenna) and spacer to the meter.
- ☐ Inspect the microwave oven, especially the door and its perimeter gasket, hinges, and latch.
- ☐ Measure the dimensions of the perimeter of the door.
- ☐ Add approximately 275 ml of tap water into the plastic or glass beaker.

- ☐ Immerse the thermometer into the water, measure and record temperature, and remove the thermometer.
- ☐ Place the beaker of water in the microwave oven and close the door.
- ☐ Turn "ON" and zero the microwave survey meter.
- ☐ Turn "ON" the microwave oven.
- ☐ Slowly scan the perimeter and the surface of the door of the microwave oven, keeping the probe cone 5 cm from the surface of the oven for about 1 min. Record measurements at various points.
- ☐ Turn "OFF" the oven, carefully remove the beaker of water, immerse the thermometer, and record temperature to confirm that there was an increase in water temperature.
- ☐ Complete a field monitoring data form (Figure 21.3), remembering to record your name, facility and location sampled, date, and microwave meter manufacturer and model.

UNIT 22

Evaluation of Airborne Extremely Low Frequency Electromagnetic Fields: Instantaneous Area Monitoring Using a Combined Electric and Magnetic Fields Meter

LEARNING OBJECTIVES

At the completion of Unit 22, including sufficient reading and studying of this and related reference material, learners will be able to correctly:

- Name, identify, and assemble the common instruments used for conducting instantaneous monitoring of extremely low frequency (ELF) electromagnetic fields.
- Summarize the principles of monitoring for ELF electromagnetic fields.
- Conduct instantaneous measurement of ELF electromagnetic fields.
- Record all applicable sampling data using field monitoring data forms.

OVERVIEW

Occupational and even nonoccupational exposures to ELF electromagnetic fields have received increased attention. The ELF electromagnetic fields consist of electric fields and magnetic fields. Electric fields are created by the presence of electric charges associated with operating and nonoperating electrical appliances and machinery connected to an electrical power source. Magnetic fields also are associated with electrical devices, but are generated in combination with electric fields when the devices are operating and there is electric current present.

Although there is not a specific demarcation, ELF electromagnetic fields are usually below 10^3 Hz (1 kHz). Major sources of ELF electromagnetic fields are electrical transmission lines (50 or 60 Hz) and numerous electrical devices present in the occupational and nonoccupational environments. Occupations requiring installation and maintenance of electrical power lines and devices present some obvious potential sources of ELF electromagnetic fields. Relatively recently common office sources, such as office computers and visual display terminals (VDTs), have received more attention relative to ELF electromagnetic fields. Accordingly, monitoring is becoming more common to evaluate the levels of ELF electromagnetic fields generated by equipment in office settings.

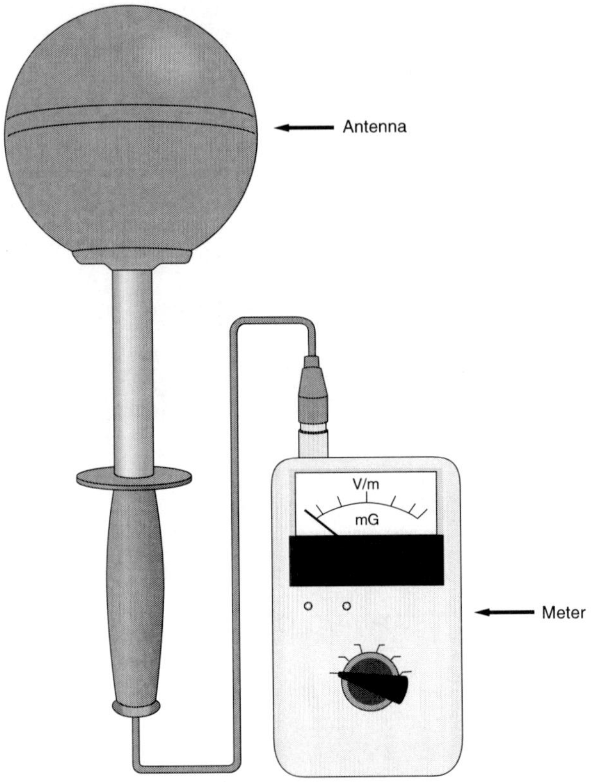

Figure 22.1 Combined electric and magnetic fields meter.

MONITORING

Monitoring requires the use of specialized electronic meters that can detect and measure ELF electric and magnetic fields (Figure 22.1). The root mean-square (RMS) of the detected signal is computed and the measurement is displayed on a direct readout meter. Magnetic fields are commonly measured (in terms of magnetic flux density) in units of milligauss (mG) or microtesla (μT) and electric fields in units of volt/meter (V/m).

Distances and heights relative to the sources of ELF electromagnetic fields must be considered during monitoring. Measurements are recorded at various vertical and horizontal distances from the source, such as a computer and VDT. In addition, it is useful to conduct measurements when the suspected sources are turned "OFF" and then when turned "ON." The contribution of electric field and magnetic field energies to the area then can be attributed to the operating source.

UNIT 22 EXERCISE

OVERVIEW

The exercise will provide the fundamental concepts for conducting instantaneous area monitoring of ELF electromagnetic fields, specifically electric and magnetic field energies, generated from a VDT using a survey meter. Place a check mark (✔) in the open box (☐) when you have obtained applicable material and completed the steps for the monitoring method.

MATERIAL

1. Monitoring for ELF Electromagnetic Fields

- ☐ Electronic ELF electromagnetic fields survey meter
- ☐ Tape measure
- ☐ Computer workstation with VDT
- ☐ Field monitoring data form (Figure 22.2)

Field Monitoring Data Form:
Real-Time Monitoring for ELF Electromagnetic Fields
Using a Combined Electric and Magnetic Field Meter

Facility Name and Location: ______________________________

Monitoring Conducted By: ______________________________

Date Monitoring Conducted: ____________

Monitoring Instrument (Type/Manufacturer/Model): ______________________

____

Source: ______________________ **Area:** ______________________

Distance (cm)	Height (cm)	Electric Field (Volts/m)	Magnetic Field (mGauss)	Source/Area Diagram and Sample Points

Field Notes:

____

____

Figure 22.2 Field monitoring data form of ELF electromagnetic fields measurements.

METHOD

1. Monitoring for ELF Electromagnetic Fields

- ☐ Refer to the survey meter manufacturer's instructions for assembly and operation of the instrument. Note that the meter is factory calibrated and must be returned to the manufacturer or the equivalent for periodic adjustments and recalibration.
- ☐ Measure the dimensions of the area.
- ☐ Turn "ON" and zero the ELF electromagnetic fields survey meter.
- ☐ Assure that the computer and VDT (and other peripherals such as printer) are turned "OFF."
- ☐ Using the tape measure, record levels of ELF electromagnetic fields at various distances from the computer and VDT screen, specifically including distances of 5 cm, 25 cm, 50 cm, and 1 m and heights of 50 cm, 1 m, 1.5 m, and 2 m.
- ☐ Turn "ON" the computer and VDT and repeat measurements.
- ☐ Complete a field monitoring data form, remembering to record your name, facility and location sampled, date, and EMF meter manufacturer and model.

UNIT 23

Evaluation of Airborne Ionizing Radiation: Instantaneous Area Monitoring Using an Ionizing Radiation Meter

LEARNING OBJECTIVES

At the completion of Unit 23, including sufficient reading and studying of this and related reference material, learners will be able to correctly:

- Name and identify the common instruments used for conducting instantaneous area monitoring of ionizing radiation.
- Summarize the principles of monitoring for ionizing radiation.
- Conduct instantaneous area monitoring of ionizing radiation.
- Record all applicable sampling data using field monitoring data forms.

OVERVIEW

The amount and rate of exposure to ionizing radiation is an important consideration in many occupational environments. These environments include clinical and research settings where radioisotopes are handled and x-ray machines are present, and nuclear power plants, and sites where radioisotopes have been stored and disposed.

Health physicists are typically involved with evaluating occupational and nonoccupational environments to detect and measure levels of ionizing radiation. Industrial hygienists are often involved directly or indirectly with radiation monitoring activities, depending on the occupational setting, the monitoring application, and the sources and levels of ionizing radiation present.

Levels of ionizing radiation that are detected and measured will vary depending on several factors, including the source and distance. As a result, area monitoring measurements can be recorded at various distances from identified or suspected ionizing radiation sources to establish a monitoring contour.

MONITORING

Instruments are available for collecting personal and area ionizing radiation data. Instantaneous area monitoring devices commonly consist of a probe connected to a meter with a direct readout display (Figure 23.1). Selection of monitoring devices depends on the type and level of ionizing

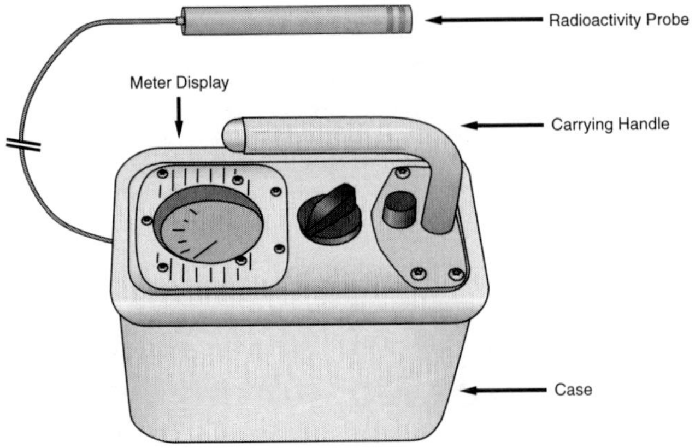

Figure 23.1 Geiger-Mueller meter.

radiation that is present. A summary of some common devices for instantaneous area monitoring for ionizing radiation follow. In addition, some personal monitoring devices also are discussed.

(i) Ionization Chamber

An ionization chamber is a real-time electronic monitoring device that consists of a chamber filled with a known volume of gas. Within the chamber is a positively charged anode and negatively charged cathode. When airborne ionizing radiation enters the chamber, it interacts with the gas molecules and forms primary ion pairs. The ions are attracted to the respective electrodes, creating a change in voltage. The change in voltage and accompanying current resulting from the ions is proportional to the amount of ionizing radiation detected. The measurement of ionizing radiation is displayed on a direct readout. Ionization chambers are useful for area measurements of relatively high levels of ionizing radiation. The meters do not differentiate between or identify the types of ionizing radiation detected.

(ii) Proportional Counters

Proportional counters are electronic meters that function very similarly as ionization chambers and are used for real-time area measurements of ionizing radiation. They are a special type of ionization chamber. The major difference is that these meters cause a magnification of ionization when ionizing radiation interacts with the gas molecules in the chamber. A higher voltage is generated and produces secondary ions. As a result of secondary ion formation, the current is amplified and corresponds to increased sensitivity of the instrument. Accordingly, proportional counters respond to lower levels of ionizing radiation than standard ionization chambers. The instrument is especially useful for detecting, measuring, and identifying alpha and beta particles.

(iii) Geiger-Mueller Meters

Geiger-Mueller (G-M) meters also operate based on the principle of ionization as discussed for the preceding two devices. The instrument exhibits even greater secondary ionization than described for the proportional counter. Accordingly, the sensitivity to detect low-level ionizing radiation is even greater. Unfortunately, the electrical current that is generated as a result of the interaction between the detected ionizing radiation and the gas molecules within the chamber and the secondary ions formed and attracted to the electrodes is not typically proportional to the ionization energy present. The G-M meter is most useful as a screening device to detect low levels of ionizing radiation, but it is not as effective for measuring accurate doses. High-level radiation can cause the meter to block, resulting in no response when levels of ionizing radiation are actually present. A major application is for area measurements of low levels of beta and gamma radiation.

(iv) Film Badges

Film badges are personal monitoring devices for collecting integrated monitoring data. The device consists of photographic film coated with a thin layer of a special emulsion containing silver halide (i.e., silver bromide). The film is positioned within a special badge that can be modified depending on the use. Interaction with ionizing radiation causes the silver halide to ionize, forming silver cations. The silver cations are attracted to negatively charged regions of the film yielding free-silver compounds. After use, the film is removed from the badge and submitted for processing. Development of the film rinses away silver halide and silver ions. Free silver, however, remains and its presence causes the film to exhibit increased opacity in proportion to the amount of ionizing radiation exposure. The measurement of exposure is determined using a densitometer that measures the degree of film opacity.

(v) Dosimeters

Pocket dosimeters are integrated monitoring devices for determining personal exposures to ionizing radiation. Like the film badge, the devices are typically clipped to a worker's lapel or shirt pocket. They are similar to ionization chamber technology. The devices reveal accumulated exposure based on a direct readout visible in the dosimeter. A similar type of pen-like dosimeter requires a separate readout device to determine the measurement of absorbed dose.

Another common personal dosimeter for detecting and measuring ionizing radiation is the thermoluminescence dosimeter (TLD). Energized electrons resulting from ionizing radiation interact and are trapped within a phosphor crystal. Thermal analysis of the dosimeter causes generation of detectable light. The measured light emission reflects the level of absorbed dose.

UNIT 23 EXERCISE

OVERVIEW

The exercise will provide the fundamental concepts for conducting instantaneous area monitoring of low-level ionizing radiation generated from a known source with known activity using an applicable electronic survey meter for ionizing radiation, such as a G-M meter. Place a check mark (✔) in the open box (☐) when you have obtained applicable material and completed the steps for the monitoring method.

MATERIAL

1. Monitoring for Low-Level Ionizing Radiation

- ☐ Electronic survey meter (e.g., G-M meter)
- ☐ Tape measure
- ☐ Commercially available known sources of low-level radiation with known activities detectable and measurable by the instrument in use
- ☐ Field monitoring data form (Figure 23.2)

Field Monitoring Data Form:
Real-Time Monitoring for Ionizing Radiation
Using a Geiger-Mueller Meter

Facility Name and Location: ______________________________

Monitoring Conducted By: ______________________________

Date Monitoring Conducted: ______________

Monitoring Instrument (Type/Manufacturer/Model): ______________________________

____

Source: ______________________ **Area:** ______________________

Distance (cm)	Height (cm)	Level (r/hr)	Source/Area Diagram and Sample Points

Field Notes:

____

____

Figure 23.2 Field monitoring data form for ionizing radiation measurements.

METHOD

1. Monitoring for Low-Level Ionizing Radiation

- ☐ Refer to the survey meter manufacturer's instructions for assembly and operation of the instrument. Note that the meter is factory calibrated and must be returned to the manufacturer or the equivalent for periodic adjustments and recalibration. A low-level reference source may be present for field checks.
- ☐ Measure the dimensions of the area.
- ☐ Assign someone to hide the known sources of low-level ionizing radiation of various locations within the area.
- ☐ Turn "ON" and zero the survey meter.
- ☐ While holding the instrument, walk around the area, listen for the audible signal of the instrument, and look at the readout to determine the measurement.
- ☐ Move closer to an area as the meter reading and the audible sound increase and identity the location of the hidden source.
- ☐ Use the tape measure and record levels at various distances from the source.
- ☐ Complete a field monitoring data form, remembering to record your name, facility and location sampled, date, and ionizing radiation meter type, manufacturer, and model.

UNIT 24

Evaluation of Ergonomic Factors: Conducting Anthropometric and Workstation Measurements

LEARNING OBJECTIVES

At the completion of Unit 24, including sufficient reading and studying of this and related reference material, learners will be able to correctly:

- Identify methods to determine whether ergonomic health and safety hazards exist in an occupational environment.
- List major equipment and materials used to collect ergonomic data.
- Describe the steps involved in performing an ergonomic task analysis.
- Collect anthropometric measurements.
- Evaluate a computer workstation.

OVERVIEW

Examination of the interface between people and workplaces within the occupational environment can provide the clues necessary to identify potential ergonomic problems. Data obtained from the Occupational Safety and Health Administration (OSHA) 200/300 logs; medical histories of work-related musculoskeletal injuries; and trends in absenteeism, complaints, and grievances may be correlated with job classifications and work locations to assist in focusing efforts to identify jobs and job tasks that exhibit ergonomic risk factors. The primary risk factors associated with the development of work-related musculoskeletal disorders include extreme or awkward postures, excessive force or exertion, frequent or repetitive movements, contact stress between the worker's body and their workplace, vibration, and extreme temperatures. These environmental demands of working may then be compared to existing data on the capabilities of workers to determine where there is a mismatch. When the environmental demands exceed the capabilities of the worker, then accidents and injuries will occur. Data on the environmental demands may be obtained by direct observation of the worker performing his job. Videotaping of the job allows a more thorough ergonomic assessment to be completed at a later time. It also provides a permanent record of the manner in which the job is performed. The ergonomic assessment may be completed on existing jobs or when there is the introduction of new processes, equipment, or materials, which may have introduced new ergonomic hazards.

EVALUATION

The first step in completing an ergonomic assessment is to determine jobs that exhibit the greatest likelihood of causing work-related musculoskeletal disorders, using the data previously described above. Then a task analysis should be performed in which the job is broken down into discrete steps or tasks that the worker must perform. For each job task, determine the postures assumed by the worker, the magnitude of the forces that the worker is exerting, the frequency and duration of performing each task, the physiological requirements of the task, vibration, and lighting. These data may be put into a chart form in which the posture, force, frequency, duration, and physiological data are described vertically and the job tasks are placed horizontally[1] (Figures 24.1 and 24.2). The variables in the vertical columns for each task are then compared to existing normative data on measures of joint ranges of motion, anthropometric data (e.g., body segment lengths and reaches), muscle strength (e.g., pulling/pushing, grip forces), information processing, and visual capabilities of workers[2]. Any task that exceeds the normative measures should be considered a task that needs to have the environmental demands reduced, eliminated, or redesigned.

The equipment needed to measure the extent of the risk factors may include, but are not limited to, video cameras, tape measures, rulers, strain/force measuring gauges, calipers, portable physiological indices monitors (e.g., heart rate monitors or galvanic skin response instruments), and stopwatches. The physical dimensions of the workplace/workstation, tools and equipment should also be measured to determine heights, reaches, depths, etc.

Worker perception surveys and psychophysical data should also be obtained. Perception surveys may be used to solicit information from workers on the difficult aspects of the job and what types

Task Analysis Worksheet

Worker Job Title:					
Job Tasks	**1**	**2**	**3**	**4**	**5**
Task Frequency					
Back • **Flexion** • **Extension** • **Rotation**					
Neck • **Flexion** • **Extension** • **Rotation**					
Continue with other Body Segments					
Forces • **Push/Pull** • **Grip**					
Lighting					
Vibration					

Figure 24.1 Sample task analysis of a job. (Adapted from Drury, C.G., A biomechanical evaluation of the repetitive motion injury potential of industrial jobs, *Semin. Occup. Med.*, 2(1), 41–49, 1987.)

[1] Drury, C.G., A biomechanical evaluation of the repetitive motion injury potential of industrial jobs, *Semin. Occup. Med.*, 2(1), 41–49, 1987.

[2] Rodgers, S.H., *Ergonomic Design for People at Work*, Vol. 2, Van Nostrand Reinhold, New York, 1986.

Field Evaluation Data Form:
Personal Anthropometric Measurements

Identification of Subject	Sex	Age (yr)	Wt. (kg)	Ht. (cm)	Sit: Eye Height (cm)	Sit: Shoulder Height (cm)	Sit: Elbow Height to Floor (cm)	Sit: Elbow Height to Seatpan (cm)	Sit: Wrist Height to Floor (cm)	Sit: Thigh Length (cm)	Sit: Popliteal Height (cm)

Worker Complaints/ Previous Injuries:

Figure 24.2 Field evaluation form for anthropometric measurements.

of changes they perceive could eliminate or reduce the difficult job tasks. Various psychophysical scales have been developed to provide subjective data on the worker's perception of job exertion, comfort, and location of any discomforts.

Ultimately, ergonomic hazards can be eliminated or minimized by modifying the tools or workstation, changing the process, reducing the frequency or repetitive motions, and modifying the methods for completing the task (Figure 24.3). It is also essential to obtain worker input into the redesign process, since they know the intricacies of their job which may not be readily evident in a quick ergonomic assessment; this will also ensure worker "buy-in" to use the redesign once it has been implemented.

UNIT 24 EXERCISE 1

OVERVIEW

This exercise will provide the fundamental concepts for conducting an ergonomic evaluation involving anthropometric measurements for a computer workstation. Record the measurements for at least ten subjects. Place a check mark (✔) in the open box (☐) when you have obtained applicable material and completed the steps for the evaluation method.

MATERIAL

1. Conducting Anthropometric Measurements and Task Analysis

☐ Tape measure
☐ Field evaluation form (Figures 24.1 and 24.2)

METHOD

1. Conducting Anthropometric Measurements and Task Analysis

☐ Identify at least ten subjects for measurements.
☐ Measure eye height while sitting.
☐ Measure shoulder height while sitting.
☐ Measure elbow rest height from the floor while sitting.
☐ Measure elbow rest height from chair seat pan while sitting.
☐ Measure wrist/hand height from the floor while sitting.
☐ Measure thigh length from buttocks to popliteal area while sitting.
☐ Measure popliteal height from floor while sitting.
☐ Calculate the statistical mean, standard deviation, 5th, 50th, and 95th percentiles for each anthropometric measure.
☐ Compare the anthropometric data among the ten subjects to published data for the general population.
☐ Observe and record what the workers are doing and how they are doing their work.
☐ Record any worker complaints about their workstation.
☐ Obtain from the worker any information on previous injuries or medical conditions associated with their workstation (such as previous symptoms of carpal tunnel syndrome or tendonitis).

Field Evaluation Data Form: Workstation Measurements

Facility Name and Location: ______________________
Evaluation Conducted By: ______________________ **Date Conducted:** ____________
Workstation Occupant: ____________ **Area:** ____________

Parameter Evaluated	Measurement/ Comment	Workstation/Area Diagram
Illumination (Lux)		
VDT Characteristics		
Monitor Height/Angle Adjustable		
Glare on Monitor		
Monitor Screen Tilt		
Upper Edge Monitor Height – Floor (cm)		
Monitor Screen Distance to Worker's Eye (cm)		
Document Holder Available		
Telephone Headset Available		
Work Surface		
Number of Operators per Workstation		
Keyboard Height – Floor (cm)		
Keyboard Height Adjustability/Slope		
Mouse Location/Height		
Wrist Rest Available		
Adequate Leg Clearance		
Area of Work Surface Adequate		
Chair		
5-Arm Base		
Castors for easy ingress/egress		
Seat Pan Height – Floor		
Seat Pan Width		
Seat Pan Depth		
Seat Padding		
Seat Slopped Downwards in Front		
Armrest Height/Adjustability		
Chair Backrest Available		
Chair Backrest Adjustable Height/Angle		
Feet Supported by Floor		

Figure 24.3 Field evaluation form for workstation measurements.

UNIT 24 EXERCISE 2

OVERVIEW

This exercise will provide the fundamental concepts for conducting an ergonomic evaluation of workplace design involving measurements of a computer workstation. Place a check mark (✔) in the open box (☐) when you have obtained applicable material and completed the steps for the evaluation method. The goal is to have the subjects seated in a neutral position with the eyes facing forward with the neck at a neutral posture, the hands located on the keyboard with the hands at the same level as the elbow or slightly below, the thighs parallel to the ground, the lower legs at a 90° angle to the ground, and the feet supported on the ground or on a footrest. Workers who wear bifocal glasses will require that their video display terminal (VDT) be lower than a worker with uncorrected vision.

MATERIAL

1. Conducting Measurements of a Computer Workstation

☐ Tape measure
☐ Light meter
☐ Field evaluation form (Figure 24.3)

METHOD

1. Conducting Illumination Measurements at a Computer Workstation

☐ Measure the level of illumination at the workstation area.
☐ Measure illumination at reading surfaces.
☐ Determine if overhead lights are parallel and not in front or behind the workstation.
☐ Determine sources of artificial light.

2. Evaluating Furniture and Conducting Measurements of Distances and Heights at a Computer Workstation

(a) VDT Considerations

☐ Determine if the monitor height is adjustable.
☐ Determine if there is glare on the screen.
☐ Determine if the monitor can be tilted.
☐ Measure the upper edge of the monitor height from the floor.
☐ Measure the screen distance from the worker's eyes.
☐ Determine if a document holder is being used and it is at the same level as the VDT.
☐ If the worker uses the telephone simultaneously with computer inputting, then a telephone headset should be made available.

(b) Working Surface

☐ Determine if more than one computer operator will use the workstation.
☐ Determine if the keyboard height is adjustable.

- ☐ Measure the height of the keyboard from the floor. Ensure that the keyboard and table are adjusted in a position that permits operators to keep their elbows at their sides with their forearms extended to form an angle with the elbow of no more than 90°.
- ☐ Determine if the keyboard and table are adjustable and if the slope of the keyboard permits the operator's wrists to be supported and in a straight and neutral position.
- ☐ Measure the height of the mouse above the floor and the location of the mouse relative to the keyboard. Ensure that the mouse is within easy reaching distance and at the same level as the keyboard with the hand and wrist in a neutral position.
- ☐ Check if there is a wrist rest available.
- ☐ Determine if there is adequate leg clearance beneath the working surface, both in thigh clearance and leg extension.
- ☐ Measure the area of the work surface and ensure that there is adequate room for both computer and writing tasks.

(c) Chair

- ☐ Check if the chair has a five-arm base.
- ☐ Determine if the chair castors permit easy ingress/egress.
- ☐ Measure the height of the chair seat pan from the floor and ensure that the worker keeps his or her knees at the same height as their hips, with the thighs parallel to the ground when the arms, wrists, and hands are in the correct position at the workstation.
- ☐ Determine if the chair seat pan height is adjustable and it is adjusted to the popliteal height of the worker.
- ☐ Measure the chair seat pan width and ensure that it coincides with the buttock width of the worker.
- ☐ Measure the seat pan depth and ensure that when the back is against the backrest, the seat pan supports the thighs without cutting off circulation in the popliteal area.
- ☐ Determine if the seat is padded and that the padding is firm and non-slippery.
- ☐ Determine if the seat front is rounded downward.
- ☐ Determine if there are armrests and if the heights are adjustable. Ensure that the armrests are adjusted to the seated elbow height of the operator.
- ☐ Determine if the chair has a backrest.
- ☐ Determine if the backrest is adjustable for height, and angle it with the lumbar support fitted to the worker. The backrest angle should permit at least a 90° angle between the back and the thigh.
- ☐ Determine if the computer operator's feet are placed flat and firmly on the floor. If this is not possible, provide a footrest.

UNIT 25

Evaluation of Air Pressure, Velocity, and Flow Rate: Instantaneous Monitoring of a Ventilation System Using a Pitot Tube with Manometer and a Velometer

LEARNING OBJECTIVES

At the completion of Unit 25, including sufficient reading and studying of this and related reference material, learners will be able to correctly:

- Name, identify, and assemble the common instruments used for conducting instantaneous monitoring of ventilation system velocity pressure, velocity, and flow rate.
- Summarize the principles of monitoring of ventilation system velocity pressure, velocity, and flow rate using a pitot tube and inclined manometer assembly and a swinging vane anemometer or thermal velometer.
- Conduct instantaneous measurements of velocity pressure and capture velocity using a pitot tube and inclined manometer assembly and a swinging vane anemometer, respectively.
- Conduct applicable calculations and conversions relative to air velocity pressure, air velocity, airflow rate, hood face velocity, and hood capture velocity.
- Record all applicable sampling data using field monitoring data forms.

OVERVIEW

Engineering controls, such as local and general ventilation systems in occupational and non-occupational environments, must be evaluated periodically to assure that they are operating efficiently, providing the degree of ventilation as per design specifications, and meeting applicable agency standards and guidelines. Monitoring is conducted to record measurements inside and outside the systems. Inside measurements are taken within ducts to determine static, velocity, and total pressures. In turn, air transport velocities and flow within the system can be calculated. Measurements outside the system include face and capture velocity measurements. Face velocity refers to the movement of air in feet per minute (f/min) at the immediate opening of a hood, vent, or diffuser. Capture velocity is a measurement of velocity at a specific distance from the source of contaminant generation to the face of the hood.

MONITORING

Ventilation systems can be qualitatively and quantitatively evaluated. Qualitative evaluation can be as simple as squeezing an aspirator bulb that is connected to a smoke tube (e.g., titanium tetrachloride) and observing the movement of the smoke as it is released into and transported through the air. This simple method can be used to determine if there is airflow flowing into a room through a diffuser or out of a room through a vent or hood. In addition, observation of the movement of the smoke can assist in determining if there is more laminar vs. turbulent airflow.

Quantitative evaluation of ventilation systems is more involved. To quantitatively measure air pressures such as velocity pressure, static pressure, or total pressure within a ventilation system, the most common devices used are the pitot tube connected to an inclined fluid manometer. A pitot tube is a concentric tube that is designed to simultaneously account for total and static pressures and, via differential pressure, measure velocity pressure. The inclined manometer contains a fluid such as water or oil. Electronic manometers are one of a few alternative devices to use instead of an inclined manometer. The pitot tube is connected to the manometer before insertion into a duct (Figure 25.1). Air pressures within the duct cause the fluid in the manometer to deflect. The deflection is often measured in inches of water (inH_2O). Several velocity pressure (VP) measurements are taken horizontally and vertically across the cross-sectional area of the duct lumen (Figure 25.2). In turn, the velocity pressures are converted to velocity (V) and averaged to calculate average transport velocity in feet per minute. Average velocity multiplied by the cross-sectional area of the duct yields volumetric flow rate in cubic feet per minute (ft^3/min). A brief summary follows.

1. Monitoring Air Velocity Pressure, Velocity, and Flow Rate in a Round Duct

- Assemble the pitot tube and inclined manometer, determine cross-sectional vertical and horizontal traverse points, insert pitot tube into pre-drilled hole, and record velocity pressure measurements at each predetermined point.

2. Calculation of Average Air Velocity Pressure, Velocity, and Flow Rate

- Convert velocity pressure to velocity (assuming 70°F and 760 mmHg).

$$V\,(\text{ft}/\text{min}) = 4005 \times \sqrt{VP\,(\text{in}H_2O)} \tag{25.1}$$

- Calculate the average transport or duct velocity.

$$V_{avg} = \frac{\Sigma V}{N} \tag{25.2}$$

- Calculate the cross-sectional area ($A = \pi d^2/4$) of the duct and, in turn, the airflow rate (Q) through the duct.

$$Q\,(\text{ft}^3/\text{min}) = V\,(\text{ft}/\text{min}) \times A\,(\text{ft}^2) \tag{25.3}$$

Measurements of face and capture velocities outside of a ventilation system are commonly determined using a swinging vane anemometer or a thermal velometer (Figure 25.3). The swinging vane anemometer is a mechanical device that involves a pendulum mechanism that deflects in response to moving air. The degree of deflection is related to the velocity of the air. A readout indicates the quantitative velocity. Thermal velometers are battery operated and consist of a probe containing a thin wire that is heated to a stable temperature. Movement of air across the wire will

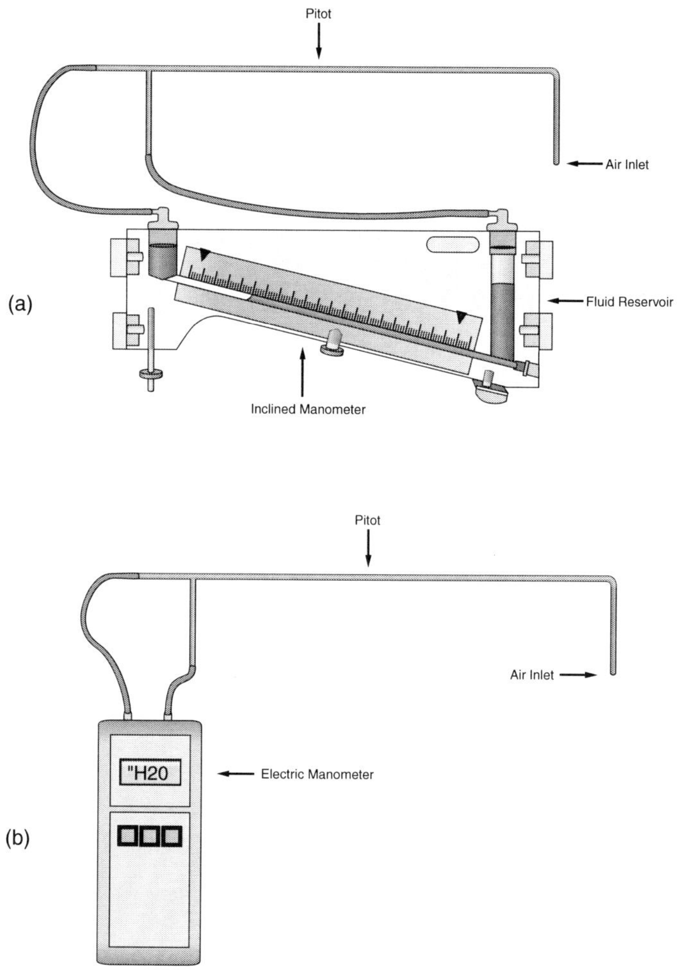

Figure 25.1 (a) Pitot tube connected to a fluid-filled inclined manometer. (b) Pitot tube connected to an electronic manometer.

cool it. The degree of cooling is related to the air velocity shown quantitatively on the meter readout. Velocity measurements are taken at several points at the face of an exhaust hood (e.g., lab hood), exhaust vent, or makeup air inflow diffuser. The measurements for the points are averaged to determine the average face velocity in feet per minute. For capture velocity, several measurements are taken at a distance from the source to the exhaust hood opening and averaged to give average capture velocity also in units of feet per minute (Figure 25.4). A brief summary follows.

3. Monitoring Capture and Face Velocity in a Laboratory Hood

- Assemble the swinging vane anemometer or thermal wire velometer, and determine cross-sectional traverse points (N) at either the sash opening or internal hood slot or duct opening. Measure and record velocities (V) at the hood sash opening for calculating the average capture velocity, or measure and record velocities within the hood along the slot or across duct opening to calculate the average face velocity.

4. Calculation of Average Air Velocity

- Calculate the average capture or face velocities as shown in Equation 25.2.

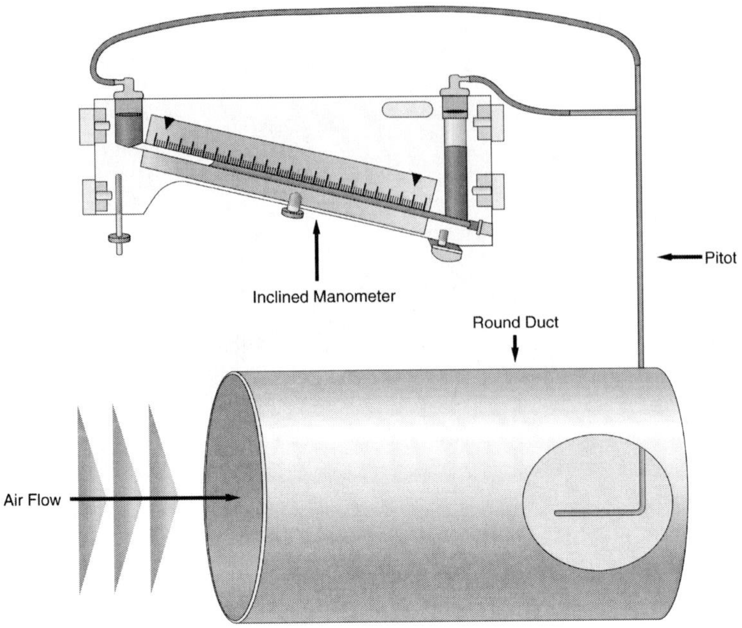

Figure 25.2 Pitot tube and manometer to measure velocity pressures at designated cross-sectional traverse points in a round duct.

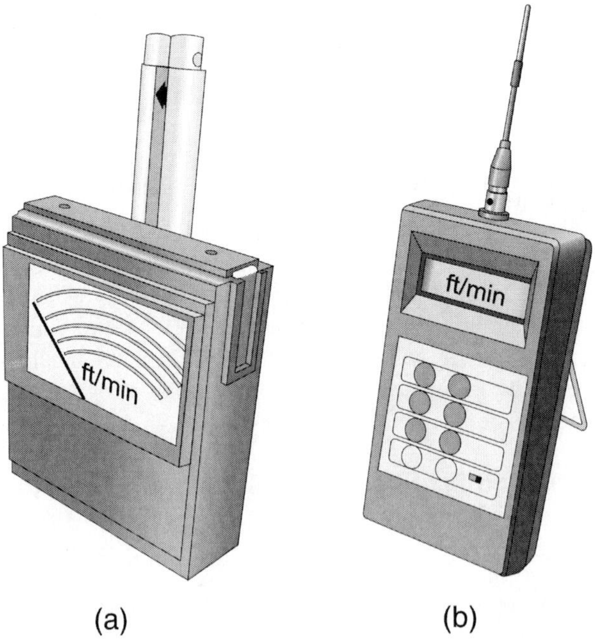

Figure 25.3 (a) Swinging-vane anemometer and (b) thermal or hot-wire velometer.

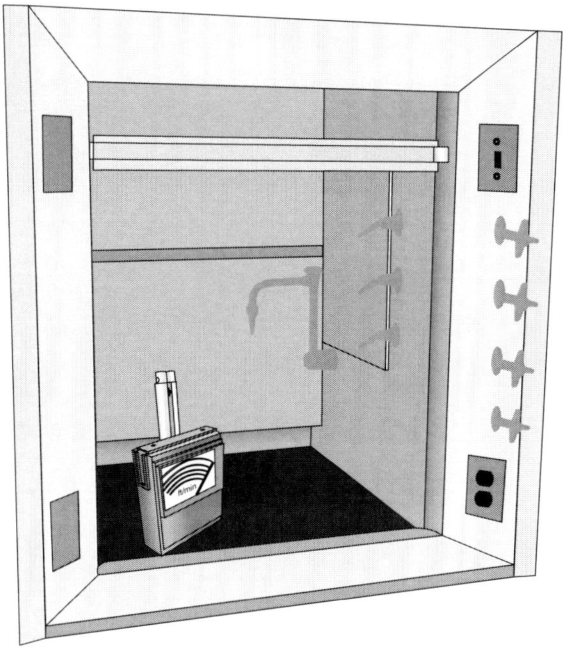

Figure 25.4 Swinging vane anemometer to measure capture velocities at designated cross-sectional traverse points in a booth hood.

UNIT 25 EXERCISE 1

OVERVIEW

The exercise will provide the fundamental concepts for conducting instantaneous monitoring of air velocity pressure, velocity, and flow through a ventilation system duct using a pitot tube and inclined manometer. Place a check mark (✔) in the open box (☐) when you have obtained applicable material and completed the steps for monitoring.

MATERIAL

1. Measurement of Duct Dimensions and Establishing Traverse Points

☐ Wax pencil
☐ Tape measure
☐ Small diameter rod at least as long as the diameter (round duct) of the duct
☐ Accessible ventilation system with round ducts
☐ Drill with 1/4-in. diameter drill bit
☐ Duct tape
☐ Field monitoring data form (Figure 25.5)

2. Monitoring Air Pressures and Determining Airflow Rate in a Ventilation System

☐ Standard pitot tube
☐ Inclined manometer with rubber hoses
☐ Manometer fluid

Ventilation Field Monitoring: Velocity Pressure, Velocity and Flow Rate

Facility: ______________________________ Date: _______________

Room: ________________ System and Location: __________________

Air Temperature = ____°F Air Pressure = _____mm Hg

Equipment:______________ Evaluator: _____________________________

System Diagram and Measurements of Duct Diameter, VP, and V
Duct Dimensions or Diameter = __________

Measurements of VP at Specific Traverse Points

No.	Vertical		Horizontal		Notes and Comments
	VP	V	VP	V	
1					
2					
3					
4					
5					
6					
7					
8					
9					
10					
Mean Velocity Pressure, VP = _________"H_2O (Convert to V) Mean Velocity, V_{avg} = ________ ft/min Area, A = _________ ft^2 Flow Rate, Q = _________ ft^3/min					

Figure 25.5 Field monitoring data form for velocity pressure measurements.

METHOD

1. Measurement of Duct Dimensions and Establishing Traverse Points

- ☐ Get permission to evaluate an existing ventilation system or use a model laboratory-scale system with round ducts.
- ☐ Locate a point that is several duct diameters downstream from the hood and at least four duct diameters upstream from bends or elbows.
- ☐ Mark two points 90° from each other using a wax pencil or marker.
- ☐ Drill a hole at each mark.
- ☐ Insert the rod into one of the holes to measure the diameter of the round duct.
- ☐ Refer to a reference to determine the location and number of pitot traverse points for the horizontal and vertical planes for the duct relative to the measured duct diameter. For example, a 12-in. diameter duct could have 5 horizontal and 5 vertical traverse points (10 total).
- ☐ Using the tape measure and wax pencil or marker, mark the individual traverse points on the pitot tube.

2. Measurement of Air Pressures in a Ventilation System

- ☐ Assure that the reservoir of the inclined manometer is half full.
- ☐ Adjust the base of the manometer so that the bubble indicator shows that the device is level.
- ☐ Turn both top ports (valves) so that they are vented to the atmosphere.
- ☐ Inspect rubber hoses to assure that there are no visible cracks and tears.
- ☐ Inspect the pitot tube to assure that it is not bent or clogged.
- ☐ Connect one rubber hose to the total pressure port (horizontal) and another rubber hose to the static pressure port (vertical) of the pitot tube.
- ☐ Connect the rubber hose from the static pressure port to the fluid reservoir of the inclined manometer.
- ☐ Connect the rubber hose from the total pressure port to the remaining port of the inclined manometer.
- ☐ Zero the leveled manometer by loosening the ruled plate and sliding it so that the meniscus of the manometer fluid is in-line with the zero mark of the plate; tighten the ruled plate.
- ☐ Gently blow into the pitot tube and assure that the manometer fluid deflects in a positive direction on the scale. Crimp each hose individually and repeat for each to assure that there are no indications of leaks.
- ☐ While crimping both hoses simultaneously, slowly insert the pitot tube into the duct (assure that ventilation system motor/fan is "ON") and release.
- ☐ Point the opening of the pitot tube and the static pressure arm in the opposite direction of airflow.
- ☐ Measure the velocity pressure at each horizontal and vertical traverse point.
- ☐ Record the data on the field monitoring data form.
- ☐ Convert each measured velocity pressure to velocity for each point.
- ☐ Average the measured air velocities.
- ☐ Use the average velocity (V_{avg}) and cross-sectional area of hood or duct opening to calculate air flow rate.

UNIT 25 EXERCISE 2

OVERVIEW

The exercise will provide the fundamental concepts for conducting instantaneous monitoring of capture velocity and flow into a booth-type hood using a velometer. Place a check mark (✔) in the open box (☐) when you have obtained applicable material and completed the steps for monitoring.

MATERIAL

1. Measurement of Hood Opening Dimensions and Establishing Traverse Points

- ☐ Tape measure

2. Measurement of Hood Face and Capture Velocity of a Ventilation System

- ☐ Aspirator bulb and smoke tubes
- ☐ Swinging vane anemometer or thermal velometer
- ☐ Field monitoring data form (Figure 25.6)

METHOD

1. Measurement of Hood Opening Dimensions and Establishing Grid Points

- ☐ Measure the length and width of the hood opening with the sash fully open.
- ☐ Establish an imaginary grid composed of 16 points within the sash opening.

2. Measurement of Hood Face and Capture Velocity of a Ventilation System

- ☐ Turn "ON" the motor/fan of the hood.
- ☐ Carefully break off the ends of a smoke tube and insert one end into an aspirator bulb.
- ☐ While pointing the tube toward the opening of the hood, squeeze the aspirator bulb and document the direction that the emitted smoke travels. Record observed movement of the smoke relative to relatively laminar vs. turbulent flow around the hood opening.
- ☐ Using the anemometer or velometer, measure the velocity at each point.
- ☐ Record the measured velocity on the field monitoring data form.
- ☐ Calculate average capture velocity. (Note: Some refer to the opening of the lab hood as the face, but since the source of contamination would be within the hood, the opening is referred to as the capture zone in this unit and elsewhere.)
- ☐ Repeat the above steps with the sash at various heights. Mark the position that allows a capture velocity of approximately 100 ft/min.
- ☐ Repeat with the hood empty and with common objects (i.e., stirrers, distillation apparatus) set up in the hood.
- ☐ Compare data from measurements recorded under various conditions.

UNIT 25 EXAMPLES

Example 25.1

What is the velocity of air if the measured velocity pressure equals 0.45 inH_2O?

Solution 25.1

$$V\,(\text{ft / min}) = 4005\ \times \sqrt{VP\ (\text{in}H_2O)}$$

$$= 4005\ \times \sqrt{0.45\ \text{in}H_2O}$$

$$= 2687\ \text{ft / min}$$

Ventilation Field Monitoring Data: Face and Capture Velocities

Facility: ______________________________ Date: _______________

Room: ________________ System and Location: __________________

Air Temperature = _____ °F Air Pressure = _____mm Hg

Equipment:______________ Evaluator: _____________________________

Hood, Diffuser or Vent Diagram and Dimensions
Dimensions: __________________

No.	Vel.	No.	Vel.	No.	Vel.	Notes and Comments
1		11		21		
2		12		22		
3		13		23		
4		14		24		
5		15		25		
6		16		26		
7		17		27		
8		18		28		
9		19		29		
10		20		30		

Mean Velocity, V_{avg} = _________ ft/min

Area, A = _________ ft^2

Flow Rate, Q = _________ ft^3/min

❑ Face Velocity

❑ Capture Velocity (Distance = ____ft)

Figure 25.6 Field monitoring data form for face and capture velocity measurements.

Example 25.2

Velocity pressures were measured at 5 horizontal and 5 vertical traverse points in a 12 in. diameter round duct. What was the average transport air velocity through the duct given the following velocity pressure measurements: 0.21, 0.25, 0.30, 0.27, 0.22, 0.20, 0.24, 0.31, 0.25, and 0.21 inH_2O? (Hint: First convert each velocity pressure measurement to velocity, then average the velocity.)

Solution 25.2

$$V_{avg}(\text{ft / min}) = \frac{\Sigma V}{N}$$

$$= \frac{19813}{10}$$

$$= 1981 \text{ fpm}$$

Example 25.3

Based on the data presented in Example 25.2, calculate the air flow rate.

Solution 25.3

$$Q\ (\text{ft}^3\text{ / min}) = V\ (\text{ft / min}) \times A\ \text{ft}^2$$

$$= 1981 \text{ ft / min } \times \pi(1 \text{ ft/2})^2$$

$$= 1981 \text{ ft / min} \times 0.785 \text{ ft}^2$$

$$= 1556 \text{ ft}^3\text{ / min}$$

UNIT 26

Evaluation of Personal Protective Equipment: Selection, Maintenance, and Fit of Dermal and Respiratory Protective Devices

LEARNING OBJECTIVES

At the completion of Unit 26, including sufficient reading and studying of this and related reference material, learners will be able to correctly:

- Describe major factors that must be considered for the selection of personal protective equipment (PPE).
- Identify various examples, components, and applications of miscellaneous dermal and respiratory protective equipment.
- Disassemble, clean and sanitize, and reassemble an air-purifying respirator.
- Explain the purpose for fit-testing individuals for respirators.
- Describe the methods for conducting field, qualitative, and quantitative fit-testing of individuals for respirators.
- Conduct a qualitative fit-test.

OVERVIEW

The occupational environment must be evaluated to assure that appropriate control measures, such as PPE, are implemented and maintained. PPE consists of devices worn by workers to prevent exposure to or direct contact with hazardous forces or substances. While PPE is a very visible and tangible symbol of an occupational safety and health program, it is not the professional's first choice for hazard control. Ideally, hazards in the workplace should be controlled through the use of engineering and administrative control measures. At times the preferred control measures are deemed either infeasible or impractical, and proper use of appropriate PPE is the best alternative. In addition, PPE must be used until appropriate and effective alternative controls (e.g., engineering, administrative) are implemented.

PPE can be classified according to the portion of the worker's body to be protected and the potential route of entry to be covered (e.g., ocular, respiratory, dermal). The primary function of PPE is to prevent or at least minimize agent contact with or entry into the worker's body. Indeed, PPE serves as an artificial barrier between workers and the contaminated occupational environment. Two major categories of PPE used in the occupational environment and discussed here are dermal protection and respiratory protection.

DERMAL PROTECTIVE GLOVES AND CLOTHING

The primary function of dermal protective equipment is to prevent agent contact with the skin. The skin, with a surface area of approximately 1.7 m^2, has the principal function of protecting internal organs from hazards in the environment. Direct skin contact with chemicals may have a variety of adverse effects. Depending on the properties of the agents (e.g., corrosive, lipophilic), local effects, systemic effects, or both can occur. In general, corrosives and irritants are often associated with local effects (e.g., burns, dermatitis); lipophilic (fat-soluble) chemicals that passively diffuse across dermal cell membranes are more likely to be associated with systemic effects (e.g., neurotoxicity). Several factors should be considered when selecting equipment, including chemical-resistant properties, design and construction, application, comfort, disposable vs. reusable equipment, manufacturer variability, and cost.

When considering the material selected for dermal protection, remember that impermeable is not an absolute, but instead a relative term. No one material will serve as a total barrier to all chemicals, and, in some instances for specific chemicals, protection may be 1 h or less before permeation and penetration occur. Permeation of a liquid or vapor through material is a three-step process involving the sorption of the chemical at the surface of the material, the diffusion of the chemical through the material, and the desorption of the chemical from the inside surface to the wearer (Figure 26.1). Factors influencing permeation include temperature, material thickness, solubility effect of the contaminant, and persistent permeation (once permeation starts, it will continue until the desorption step occurs). Breakthrough time (the rate of permeation through the material as measured by the elapsed time between surface contact and appearance of the chemical on the inside of the material) provides the professional with critical data to discriminate among products or to develop strategies for applicable program administration. In addition, some materials may degrade or break down when exposed to a specific chemical substance. Accordingly, consideration must be given to the degradation properties. Manufacturers of PPE, including clothing, will provide technical data for degradation and breakthrough for their product line.

RESPIRATORY PROTECTION

Respirators are one of the most frequently used pieces of PPE in the occupational environment. The three categories of respirators are air purifying, atmosphere supplying, and combination respirators (Figures 26.2 and 26.3). Two major types of face pieces are half- and full-face masks. A half-face respirator covers the wearer's face from the nose bridge to below the chin. A full-face respirator, however, extends from approximately midforehead to below the chin. As a result, in general, full-face-piece respirators provide more protection relative to half-mask units.

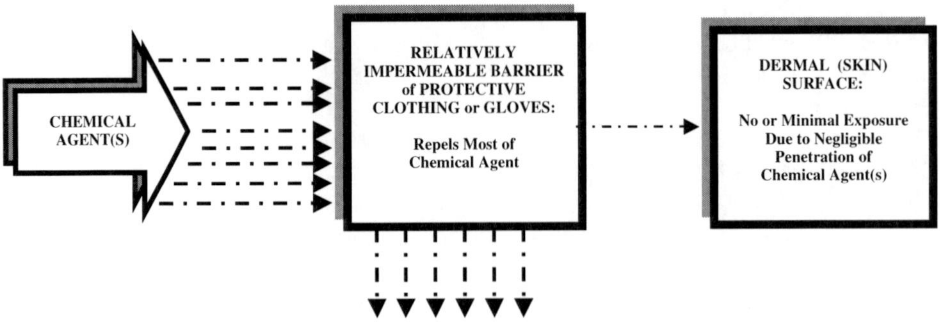

Figure 26.1 Artificial barrier for dermal protection.

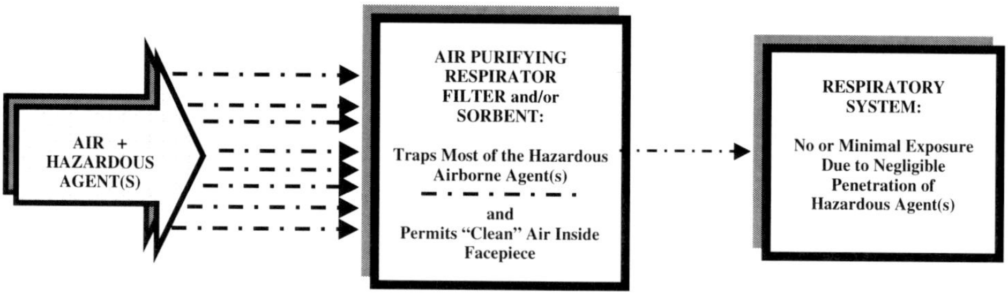

Figure 26.2 Artificial barriers for respiratory protection — air purifying respirator.

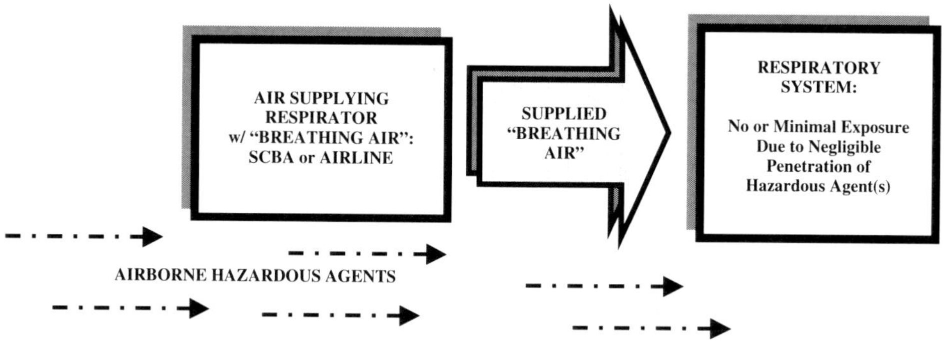

Figure 26.3 Artificial barriers for respiratory protection — air supplying respirator.

It is necessary to assure that the appropriate respirator is selected based on the application for use. It is equally important to assure that workers are evaluated for respirator fit and that the devices are cleaned and maintained. When selecting respirators, it is necessary to determine the nature of the hazards. Oxygen deficiency, physical properties of the hazard, chemical properties of the hazard, physiological effects, actual concentration of toxic compounds, occupational exposure limits, and warning properties of the contaminant must be evaluated to accurately determine exact respiratory protection requirements.

Respirators are required to be effectively cleaned and maintained. They must be continuously evaluated for defects (including leak checks). Respirators should be inspected before and after each use and at least monthly. In addition, they must be evaluated to assure that there is no evidence of mixing and matching of parts from different respirators. Finally, respirators must be stored properly to protect against dust, sunlight, heat, extreme cold, excessive moisture, and damaging chemicals. Respirators should be stored in plastic bags in cabinets to provide adequate protection from these agents and to prevent equipment distortion.

Respirators must be cleaned and disinfected after each use. Washing with detergent in warm water using a brush, thoroughly rinsing in clean water, and air drying in a clean place is acceptable. Cleaning and disinfecting should include a good detergent and sanitizer with bactericide. This can include hypochlorite solution or aqueous solution of iodine. Temperatures should be maintained between 120 and 140°F. Rinse in clean, warm water (140°F). Automatic respirator washers, similar to some degree as a dishwasher, are available as an alternative to hand washing and sanitizing respirators.

Respirators have assigned protection factors (PFs) or fit-factors based on a laboratory evaluation prior to marketing. The PF is a unitless number based on the concentration of contaminant that leaks into a respirator face-piece relative to the known concentration outside of the face-piece under test conditions.

$$PF = \frac{\text{Concentration Chemical Outside Face-piece}}{\text{Concentration Chemical Inside Face-piece}} \tag{26.1}$$

The PF, in turn, can be used to estimate the maximum use concentration (MUC) for a specific type of respirator for protection from a specific chemical agent. The MUC refers to the product of the PF multiplied by the occupational exposure limit (OEL), such as permissible exposure limit (PEL) or threshold limit value (TLV), for a given chemical agent.

$$MUC = PF \times OEL \tag{26.2}$$

The PF assigned for a specific type of respirator assumes that the respirator fits the worker and is worn properly. To assure that the respirator fits a worker, a fit-test is required. Fitting involves methods that ensure that adequate protection is afforded to respirator users (e.g., to duplicate all facial movements used in normal conversation, a speech pathologist developed what is known as the Rainbow Passage). During the fit-test, the subject either reads the paragraph aloud or is asked to articulate other speech such as counting. In addition, the subject will be asked to turn his head from side-to-side and up-and-down. There are two methods of fit-testing, qualitative and quantitative.

(i) Qualitative Fit-Test

This is a subjective evaluation of the quality of respirator fit. In a qualitative fit-test, a test atmosphere capable of stimulating the senses is generated around the wearer of a respirator equipped with the filter elements efficient in removing the test agent. If subtle leakage occurs, the wearer's senses will detect the stimulant. Three common fit-test agents are isoamyl acetate vapor (referred to as banana oil), saccharin, or irritant smoke (stannic chloride), which stimulate an involuntary smell, taste, and cough, respectively. If the fit-test qualitatively confirms that the respirator fits, then the specific protection factor assigned to the specific type of respirator can be assumed. While still an acceptable fit-testing method, the qualitative technique is becoming obsolete and in some cases unacceptable. For example, the Occupational Safety and Health Administration (OSHA) Respiratory Protection Standard 29 CFR 1910.134 requires quantitative fit-testing of respirators that have assigned protection factors greater than ten.[1]

(ii) Quantitative Fit-Test

This is an objective fit-testing procedure resulting in a numerical determination of fit. The subject wears a respirator with an internal probe which permits the determination of the relative concentration of a contaminant within the breathing zone of the respirator (Figure 26.4). Sodium chloride, corn oil mist, polyethylene glycol 400 (PEG 400), and di-2-ethylhexyl sebacate (DEHS) are agents commonly used in quantitative fit-tests. A subject wearing an appropriate respirator is placed in a chamber where a known concentration of test agent is generated by a special device. A probe inserted into the face-piece of the respirator collects a sample of air inside the face-piece. The dividend of concentration outside the face-piece divided by concentration inside the face-piece provides a quantitative protection factor.

Today, a more common device is available that does not need a chamber and simply measures the concentration of ambient particulate outside and inside the face-piece for comparison. The

[1] OSHA Respiratory Protection Standard Title 29 CFR 1910.134, 1998.

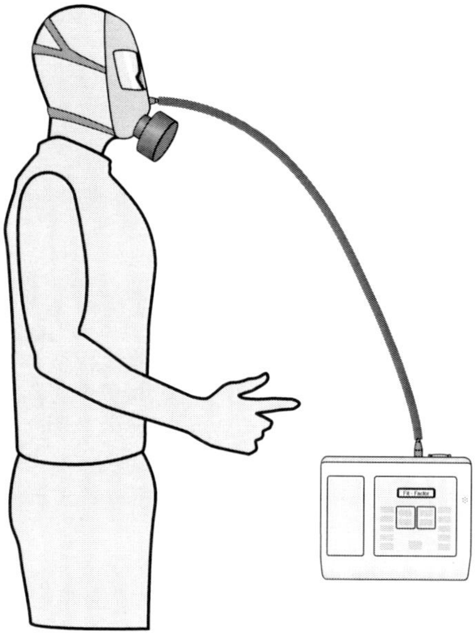

Figure 26.4 Quantitative fit-testing of full-face-piece canister-style respirator worn by subject.

technology, condensation nuclei counting, involves light scattering and detection. Ultrafine, airborne suspended particles collected from inside and outside the face-piece are first enlarged to a size that can be detected. This technology involves an instrument that collects, detects, and counts otherwise undetectable, ultrafine submicrometer (0.02 to <1 μm) particles. The sampled air containing suspended particles flows into a heated chamber where the particles mix with isopropanol vapor. The particles and vapor pass into a condenser where they cool, and the alcohol vapor condenses on the particles forming larger (>1 μm) detectable and measurable particles. Laser optics focus light on the enlarged particles causing light scattering. The scattered light is detected by a photodiode and converted into electrical pulses that are counted in direct relationship to the concentration of particles. An automatic comparison is made of inside face-piece vs. outside face-piece concentrations.

The quantitative fit-test is more accurate in determining fit, since it provides an objective assessment of the amount of leakage of contaminant outside the face-piece to the inside. Based on the results of a quantitative fit-test of a specific type of respirator for a specific wearer, a specific protection factor can be calculated and assigned. The calculated PF based on actual quantitative fit-testing is frequently higher than the PF normally assigned to the respirator. Accordingly, the PF is specific to the wearer's fit and can be used to verify the assigned PF of the selected face-piece.

(iii) Negative and Positive Pressure Check

Either one of the qualitative and quantitative (preferred) fit-test procedures summarized above is conducted prior to initial selection and initial use of a respirator. To qualitatively evaluate fit prior to each use, a simple positive-negative fit-check procedure is also required. The simple test involves using the palms of your hands to cover the inlet ports (e.g., cartridge or canister opening) while inhaling (negative pressure check) or exhalation ports when exhaling (positive pressure check). There should be no influx or efflux of air during the negative and positive pressure checks, respectively. Negative and positive pressure field tests are performed to determine the adequacy of the face-piece to face-seal.

UNIT 26 EXERCISE

OVERVIEW

The exercise will provide the fundamental concepts for identifying and selecting miscellaneous dermal and respiratory protective equipment. In addition, principles of cleaning and sanitizing respirators and conducting a qualitative fit-test procedure will be demonstrated. The learner is encouraged to review Appendix A of the OSHA Respiratory Protection Standard for more detailed procedures for both qualitative and quantitative fit-test methods. Place a check mark (✔) in the open box (☐) when you have obtained applicable material and completed the steps for the exercises.

MATERIAL

1. Identification, Selection, and Inspection of Dermal and Respiratory Protective Equipment

☐ Gloves
☐ Clothing
☐ Half-face and full-face air purifying respirators with cartridges or canisters
☐ Airline and self-contained air-supplied respirators with air cylinders

2. Cleaning and Sanitizing Respirator Face-Pieces

☐ Plastic bucket or basin with ≥2 gal capacity
☐ Cleaning solution (liquid dish soap in water) or commercially prepared cleaner for cleaning respirators
☐ Sanitizing solution (2-ml bleach per 1-l water) or commercially prepared solution for sanitizing respirators
☐ Soft bristle brush
☐ Lint-free cloth
☐ Clean, labeled, sealable plastic bag for respirator

3. Qualitative Fit-Test of an Air-Purifying Respirator

☐ Qualitative fit-test chamber consisting of a clear polyethylene bag approximately 24-in. diameter and 60-in. long with at least 4-ml thickness draped over a 24-in. diameter disk cut from plywood or heavy cardboard with a 1/4-in. diameter eyebolt approximately 2-in. long with two nuts and washers and a 2-in. threaded screw hook positioned in the center
☐ Selection of air-purifying respirators with appropriate cartridges or canister
☐ Stannic chloride smoke tubes and aspirator bulb for the irritant smoke test or isoamyl acetate ampoules for the banana oil test
☐ Respirator fit-test data form (Figure 26.5)

4. Quantitative Fit-Test of an Air-Purifying Respirator

☐ Quantitative fit-test meter (e.g., condensation nuclei counter technology)

Field Evaluation Data Form:
Qualitative and Quantitative Fit-Test of Respirator

Facility Name and Location: ______________________________

Evaluation Conducted By: ______________________________

Date Evaluation Conducted: ____________

Type of Fit-test Procedure (check one):

- ☐ **Qualitative Fit-test (Pass/Fail Result):**
  Fit-Test Medium (check one): ☐ **Smoke Tube** ☐ **Isoamyl Acetate** ☐ **Saccharin Solution**

- ☐ **Qualitative Fit-test (Fit Factor Result):**
  Instrument Manufacturer/Model: ______________________________
  Aerosol: ____________

Name	Respirator Facepiece Type	Respirator Manufacturer/ Model	Filters/ Cartridges	Respirator Size	Fit-Test Result (Pass/Fail) or (Fit Factor)

Field Notes:

Figure 26.5 Field evaluation form for qualitative and quantitative fit-test of respirators.

- ☐ Selection of air-purifying respirators with appropriate cartridges or canister (must be fitted with the appropriate sampling port and line corresponding with the fit-test meter used)
- ☐ Respirator fit-test data form (Figure 26.5)

METHOD

1. Identification and Selection of Dermal and Respiratory Protective Equipment

- ☐ Identify the different types of personal protective gloves, clothing, and respirators provided to you by the instructor.
- ☐ Select which gloves, clothing, and respirator would be acceptable for various agents and conditions presented by the instructor.

2. Inspection of Dermal Protective Equipment

- ☐ Handle and view the personal protective gloves and clothing provided to you and observe for holes, cuts, tears, cracking, deterioration, or other flaws that may compromise the efficiency of these artificial barriers.

3. Inspection of Respiratory Protective Equipment

- ☐ Check disposable air-purifying respirator for:
  - ☐ Holes in the filter or damage to the sorbent (e.g., loose granules).
  - ☐ Straps with poor elasticity or deterioration (e.g., brittleness).
  - ☐ Rusting or deteriorated metal nose clip.
- ☐ Check reusable air-purifying respirator:
  - ☐ Face-piece for dirt, pliability, deterioration, cracks, tears, or holes.
  - ☐ Straps for breaks, tears, elasticity, deterioration, and broken clips.
  - ☐ Inhalation and exhalation valves for holes, warpage, tears, and dirt.
  - ☐ Filter/sorbent cartridges and canisters for dents, corrosion, expiration dates (if applicable); and gaskets for deterioration, warpage, tears, and dirt.
  - ☐ Correct components and filters or cartridges.
- ☐ Check airline air-supplying respirators:
  - ☐ Follow checklist for air-purifying respirators as outlined above.
  - ☐ Check hood, helmet, shroud, or suit for cracks, tears, and deterioration.
  - ☐ Check face-shields for cracks or breaks, abrasions, or distortions that could interfere with vision.
  - ☐ Check air supply system for air quality, breaks/kinks in hoses, couplings, regulators/valves.
  - ☐ If a compressor is used, check for air-purifying system and CO/high temperature warning alarm.
- ☐ Check self-contained air-supplying respirators:
  - ☐ Follow checklist for air-purifying respirators as outlined above.
  - ☐ Check the integrity and air/oxygen pressure for the cylinder, regulator, harness assembly, straps, and buckles.
  - ☐ Assure that the regulator and warning alarm function.

4. Cleaning Respirator Face-Pieces

- ☐ Obtain at least one type of reusable respirator face-piece.
- ☐ Disassemble the respirator.
- ☐ Identify the major components of the respirator, including face-piece, inhalation valve, exhalation valve, connecting tube, canisters/cartridges, face lens (if applicable), straps, clamps, and corrugated breathing tubes (if applicable).
- ☐ Immerse the components of the respirator into the cleaning solution, gently brush (if necessary), and rinse in clean, warm (140°F) water.

- ☐ Immerse the components of the respirator into the sanitizing solution and rinse in clean, warm water.
- ☐ Reassemble the respirator after it has been allowed to air dry or manually dried using a lint-free cloth.
- ☐ Conduct final inspection of all components assuring proper assembly.
- ☐ Place respirator in a clean, labeled bag and seal.
- ☐ Store respirator so that it is protected from dust, air contaminants, harmful chemicals, sunlight, excessive heat and cold, and moisture. Avoid hanging respirators by their straps.

5. Positive-Negative Fit-Check of Respirator Face-Pieces

- ☐ Negative pressure check:
  - ☐ Don an air-purifying respirator with acceptable cartridges or canister secured in place.
  - ☐ Adjust the face-piece and straps so that the respirator fits comfortably, but snugly.
  - ☐ Close off the inlet of the canister, cartridge, or filter by covering with the palms or squeezing the breathing tube (if applicable).
  - ☐ Inhale so face-piece collapses and hold breath for 10 sec.
  - ☐ Fit is acceptable if the face-piece remains collapsed without signs of influx of air.
- ☐ Positive pressure check:
  - ☐ With the respirator still on, close off exhalation valve on respirator by covering with your palm.
  - ☐ Exhale into the face-piece.
  - ☐ Fit is acceptable if pressure builds without leakage based on efflux of air.

6. Qualitative Fit-Test of an Air-Purifying Respirator

- ☐ Suspend the bag from the eyebolt so the chamber ceiling is about 7 ft from the floor.
- ☐ Select a volunteer to don an air-purifying respirator with appropriate cartridges or canister. Assure the test subject has received orientation on respiratory protection and is medically approved to wear a respirator. Use cartridges or canister approved for particulate if using a test aerosol or approved for organic vapors if using the isoamyl acetate. As an alternative, use a combination cartridge or canister approved for both particulate and organic vapor.
- ☐ Explain the steps of the qualitative fit-test to the individual so that he or she understands and is comfortable with the concept and the purpose. Inform the individual that he may request to exit the chamber at any time he feels anxious or experiences discomfort.
- ☐ Assure that the respirator is positioned and secured properly. Conduct a preliminary negative-positive fit-check.
- ☐ Have the individual enter the fit-test chamber.
- ☐ If using the irritant smoke, break open both ends of the glass tube, connect the tube to the aspirator bulb, and aspirate several plumes through a hole into the fit-test chamber. Alternatively, if using isoamyl acetate, suspend an open ampoule or a saturated paper towel in the chamber.
- ☐ Request the individual to breathe normally, breathe deeply, turn head from side to side, move head up and down, read the rainbow passage, bend over at the waist, and then breathe normally.

7. Quantitative Fit-Test of an Air-Purifying Respirator

- ☐ Select a volunteer to don an air-purifying respirator with appropriate cartridges or canister. The face-piece must have the appropriate sampling port corresponding with the fit-test meter used. Use cartridges or canister approved for particulate (e.g., HEPA filter; N95).
- ☐ Explain the steps of the quantitative fit-test to the individual so that he/she understands and is comfortable with the concept and the purpose. Inform the individual that he may request to discontinue the procedure at any time he feels anxious or experiences discomfort.
- ☐ Assure that the respirator is positioned and secured properly. Conduct a preliminary negative-positive fit-check.
- ☐ Assure that the sampling line is connected from the face-piece port to the fit-test meter. Follow the fit-test meter manufacturer's instructions for operating and begin the test. Also, have the

volunteer don the respirator for 5 min before starting the actual quantitative fit-test to assure that the ambient particles potentially trapped inside the respirator are purged.

- ☐ Request the individual to breathe normally, breathe deeply, turn head from side to side, move head up and down, bend over at the waist, grimace, and then breathe normally.
- ☐ Read the display for the protection or fit factor.

UNIT 27

Evaluation of Personal Pulmonary Function: Instantaneous Monitoring Using an Integrated Electronic Spirometer

LEARNING OBJECTIVES

At the completion of Unit 27, including sufficient reading and studying of this and related reference material, learners will be able to correctly:

- Name, identify, and assemble the common instrument used for conducting instantaneous monitoring of pulmonary function.
- Summarize the principles of monitoring pulmonary function.
- Conduct instantaneous measurements of pulmonary function.
- Record all applicable sampling data using field monitoring data forms.

OVERVIEW

Pulmonary function testing is the measurement of lung volumes and performance with the purpose of identifying respiratory impairments that may not otherwise be identifiable by other methods. Pulmonary function testing is not a routine responsibility of industrial hygienists. As discussed in Unit 18 for audiometric evaluation, industrial hygienists are unlikely to have the credentials or reason to conduct pulmonary function tests (PFTs). In most cases, pulmonary function testing is conducted by a licensed, registered, or certified clinician, such as a respiratory therapist or nurse, or administered by a person who has completed an accredited course in spirometry. Industrial hygienists are often responsible for being familiar with aspects of pulmonary function testing. Two major applications of pulmonary function testing relative to industrial hygiene are evaluation of workers for approval to wear respirators and identification of occupationally related lung diseases.

Indeed, under the Occupational Safety and Health Administration (OSHA), pulmonary function testing is a required medical surveillance component for workers who must wear respirators and for some workers exposed to agents that are toxic to the respiratory system. The OSHA Respiratory Protection Standard (29 CFR 1910.134) states that workers must have a medical evaluation including PFT to determine their health status and approval to wear a respirator[1]. Other OSHA standards

[1] Occupational Safety and Health Administration Respiratory Protection Standard Title 29 CFR 1910.134, 2001.

that address PFTs include the Asbestos Standard (29 CFR 1910.1001), Cotton Dust Standard (29 CFR 1910.1043), the Benzene Standard (29 CFR 1910.1028), and Formaldehyde Standard (29 CFR 1910.1048). PFTs typically include spirometry (with post-bronchodilator responses) and measures of lung volumes and diffusing capacity. This battery of tests (with serial PFTs and trending) allows for the diagnosis of lung impairments such as obstructive lung disease from occupational asthma and interstitial lung disease caused by asbestosis.

MONITORING

Spirometers refer to the devices that are used to conduct PFTs. The instruments measure changes in respiratory volumes. Electronic integrating spirometers have become very common relative to the volumetric displacement spirometers. An electronic integrating spirometer or pulmonary function meter is an electronic device that will measure volumetric airflow generated by a subject during inhalation and exhalation. The device consists of a mouthpiece that is connected via flexible tubing to a meter. A subject, with nostrils clipped shut, places the mouthpiece into his/her mouth and is instructed to inhale and exhale on command (Figure 27.1). The airflow from the subject is electronically converted to an electrical signal that is converted to a numerical display on a direct readout. The instruments are commonly interfaced with a computer. Spirograms are graphical representations of the data showing relationships, such as volume vs. time or volume vs. flow, for the duration of a forced exhalation.

The total volume of air forcibly exhaled to exhaustion following maximal inspiratory effort is referred to as forced vital capacity (FVC). The forced exhaled volume during the first second is called forced exhaled volume (FEV_1). FVC and FEV_1 are measured in units of liters (l), and the pulmonary function meter is capable of measuring expiratory volumes in excess of 7 l. In addition, exhaled volume can be accumulated for at least 10 sec. The ratio of FEV_1 to FVC is a very useful diagnostic measurement for clinicians.

Normal values for FVC and FEV_1 vary based on the age, weight, height, gender, and race of subjects. Much error can result if the pulmonary function meter is not calibrated or operated properly. In addition, poor subject performance can generate erroneous pulmonary function data.

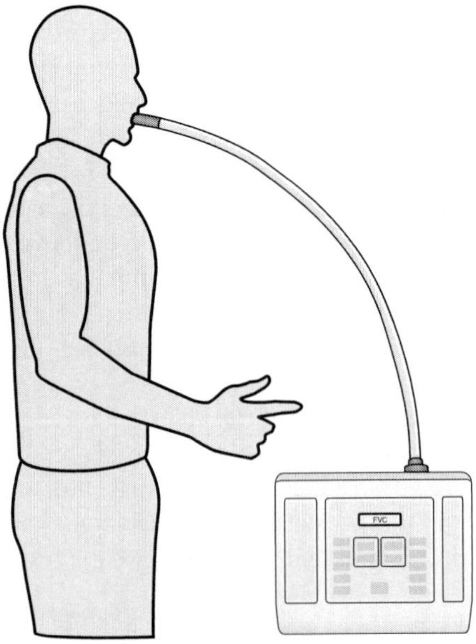

Figure 27.1 Integrated electronic spirometer or pulmonary function meter connected to subject.

UNIT 27 EXERCISE

OVERVIEW

The exercise will provide the fundamental concepts for conducting instantaneous monitoring of pulmonary function using an integrated electronic spirometer. Place a check mark (✔) in the open box (☐) when you have obtained applicable material and completed steps for the evaluation method.

MATERIAL

1. Calibration of an Integrated Electronic Spirometer

- ☐ 3-l calibration syringe

2. Conducting Pulmonary Function Measurements

- ☐ Integrated electronic spirometer or pulmonary function meter
- ☐ Disposable mouthpieces
- ☐ Nose clips
- ☐ Monitoring data form (Figure 27.2)

METHOD

1. Calibration of an Integrated Electronic Spirometer

- ☐ Read the manufacturer's instructions for assembly and operation of the pulmonary function meter. Note that the instrument is field calibrated before and after each use, but may require periodic factory calibration.
- ☐ Turn "ON" the meter.
- ☐ Retract the plunger from the calibration syringe.
- ☐ Inject 3 l of air within a maximum 10-sec period to yield positive or expired flow.
- ☐ Retract the plunger and 3 l of air within a maximum 10-sec period to yield negative or inspired flow.
- ☐ Repeat until the meter readout/computer indicates expired and inspired volumes equal to 3.0 ± 0.01 l.

2. Conducting Pulmonary Function Measurements

- ☐ Select a subject and be sure to summarize the purpose and steps of the pulmonary function test.
- ☐ Assure that the meter is turned "ON" and ready for operation.
- ☐ Attach a disposable mouthpiece to the mouthpiece connector.
- ☐ Attach the nose clip to the subject's nose to compress nostrils.
- ☐ Request the subject to place the mouthpiece into his or her mouth and form a tight seal with his lips.
- ☐ Instruct the subject to relax and breathe normally to yield a graphic tracing of tidal breathing.
- ☐ When the tracing indicates that tidal breathing is steady, instruct the subject of inhale maximally and then quickly forcibly exhale maximally as long as he can.
- ☐ Record the measurements of FVC and FEV_1.
- ☐ Repeat two more times and record the best data of the three trials.
- ☐ Complete the monitoring data form, remembering to record the technician's name, subject's name, age, weight, height, sex, race, and the integrated electronic spirometer manufacturer and model.

Field Evaluation Data Form:
Personal Pulmonary Function Test

Facility Name and Location: ______________________________

Evaluation Conducted By: ______________________________

Date Evaluation Conducted: ______________

Monitoring Equipment (Type/Manufacturer/Model): ______________________

____

Pre-Test Calibration: _____**L** **Post-Test Calibration:** _____**L**

Identification of Personnel	Gender	Age (yr)	Height (cm)	Weight (kg)	Pulmonary Function Test Result		
					FVC (L)	FEV_1 (L)	FEV_1: FVC Ratio

Comments/Notes:

Figure 27.2 Field evaluation data form for pulmonary function measurements.

APPENDIX A

Industrial Hygiene Sampling Strategies, Calculations of Time-Weighted Averages, and Statistical Analysis

LEARNING OBJECTIVES

At the completion of Appendix A, including sufficient reading and studying of this and related reference material, learners will be able to correctly:

- List the factors that must be determined when monitoring employee exposure.
- Differentiate between area and personal sampling.
- List and describe the four types of sampling options.
- Differentiate between descriptive and inferential statistics.
- Explain the differences between systematic and random error.
- Describe the application of descriptive statistics and error concepts to sampling and analytical errors (SAEs).
- Calculate upper confidence limit (UCL) and lower confidence limit (LCL) for full-period single and consecutive samples.
- Calculate time-weighted average (TWA) exposures for shortened shifts (<8 h), normal shifts (8 h), and extended shifts (>8 h).
- Determine permissible exposure limits (PELs) for compliance and noncompliance based upon calculations.

OVERVIEW

According to the American Academy of Industrial Hygiene's Code of Ethics for the professional practice of industrial hygiene, the industrial hygienist's primary responsibility is to protect the health of employees. As a consequence, one of the most important goals of any industrial hygiene program is to accurately determine employee exposure to environmental stressors. Once potential health hazards have been recognized, the industrial hygienist often monitors the workplace or worker to establish exposure concentrations. Evaluation of risk can then be determined by comparing the concentrations obtained during monitoring against recommended guidelines (e.g., threshold limit values [TLVs]) and mandatory standards (e.g., PELs). Control strategies and options can be based in part or entirely on the results obtained during monitoring and the risk assessment activity.

The ultimate success of this exposure and risk assessment process hinges upon the validity and reliability of the measurements obtained during workplace or worker monitoring. Several factors must be determined when establishing an employee exposure monitoring program.

- Which employee or employees should be sampled?
- What type of samples should be collected?
- How many samples should be collected?
- What length of sampling intervals for partial period samples?
- When should samples be collected?

Logic indicates that it is cost and time prohibitive to continuously sample all employees. One method suggests that the employees with the maximum risk or highest exposure potential should be sampled to establish exposure measurements. Where there are several locations that pose significant risk, employees should be sampled in each high risk location. An alternative approach is to perform a comprehensive exposure assessment — a systematic method for defining and judging all exposures for all workers on all days. During a comprehensive exposure assessment, the industrial hygienist systematically reviews the operations, tasks, materials, and job assignments for all workers. Since the majority of chemicals do not have a TLV or PEL, using a comprehensive exposure assessment methodology allows the industrial hygienist to move away from just sampling the maximum risk employee for compliance determination to characterizing the exposures of all workers on all days.

There are five major steps in an exposure assessment strategy:

- Basic characterization (collecting information to describe the workplace, workforce, and environmental agents)
- Exposure assessment (determining whether exposures are acceptable, unacceptable, or uncertain based on the information collected during the basic characterization)
- Further information gathering (collecting additional information to resolve uncertainties and may include exposure monitoring)
- Health hazard control (controlling unacceptable exposures)
- Reassessment (periodically reevaluating the information and exposures already collected)

By applying the reassessment step, the strategy is cyclic and is most effective when used iteratively[1,2].

The decision to take area samples (where monitoring equipment is set up in a given location to determine the concentration of the generated contaminant) or personal samples (where monitoring equipment is placed on the employee to determine individual exposures) must be based upon employee mobility issues. This is because these issues relate to the likelihood of exposure as well as the probability of obtaining accurate and representative exposure samples. Area sampling is typically performed to identify high exposure locations, determine flammable/explosive concentrations, evaluate concentrations in confined spaces, and continuously monitor for leakage in high risk locations. Personal sampling is performed when it is necessary to determine both individual exposures and compliance with standards and guidelines. If maximum risk employees cannot be determined or if a comprehensive exposure assessment is driving the monitoring, it will be necessary to implement random sampling procedures. Tables and strategies are presented in numerous publications and should be referred to if this strategy has been selected[3].

Once the decision has been made to take personal or area samples, the type of sampling strategy must be established. The industrial hygienist has four sampling options:

- Full-period single sample — One measurement is taken continuously over an entire workshift for an area or individual. An example of this sampling strategy would be attaching a respirable dust

[1] Mulhausen, J.R. and Damiano, J., *A Strategy for Assessing and Monitoring Occupational Exposures,* AIHA Press, Fairfax, VA, 1998.

[2] Conrad, R.G. and Soule, R.D., Principles of evaluating worker expsoure, in *The Occupational Environment: Its Evaluation and Control*, DiNardi, S., Ed., AIHA Press, Fairfax, VA, 1997.

[3] Leidel, N.A., Busch, K.A., and Lynch, J.R., *Occupational Exposure Sampling Strategy Manual*, DHEW (NIOSH) Publication No. 77-173, Washington, D.C., 1977.

sampling train to an employee at the beginning of the workshift and removing the monitoring equipment at the end of the 8-h workshift.

- Full-period consecutive samples — This sampling strategy requires that several samples be taken during the entire workshift. The workshift is divided into consecutive segments that are often, but not necessarily, of equal duration. Using the same example as above, the industrial hygienist can attach the respirable dust sampling train to the employee's collar at the start of the shift. After a 2-h interval has elapsed, the cassette can be replace with another cassette and so on until four 2-h samples are obtained for the employee.
- Partial-period consecutive samples — This sampling strategy employs several periods of samples taken over the course of a workshift, but some periods of the shift are not sampled. It is assumed that periods not sampled had no exposure (such as a beginning of a shift, the end of a shift, or a lunch break) or that exposures were equivalent to those sampled.
- Grab samples — Typically 5 min or less in duration, grab sampling techniques employ multiple samples (at least seven) taken at randomly selected intervals during the workshift. This strategy is usually more difficult to work with from a statistical perspective because they tend to reflect a lognormal distribution.

As part of the sampling protocol, remember that field blanks must be submitted for laboratory analysis along with the field samples. Field blanks are sampling media that have been briefly opened and then closed in the monitoring area during the sampling event; they are treated identical to field samples except no air is pulled through them. Field blanks are used to verify that the media have not been contaminated during manufacturing, shipping, and field activity or to check laboratory analytical reliability by comparing contaminant concentrations of the blanks to sample concentrations. A rule of thumb for the number of field blanks is to submit one blank for every ten field samples, with a minimum of two blanks. The specific number of field blanks to submit, however, is specified by the method and the analytical laboratory.

TIME-WEIGHTED AVERAGE CALCULATIONS

In order to determine compliance with regulations (PELs) and guidelines (TLVs) for worker exposures, the averaging time must be the same. The standard workshift is 8 h, so the PELs and TLVs are expressed as 8-h TWAs. In order to determine compliance, worker exposures must also be reported as 8-h TWAs.

When sampling is conducted during the entire 8-hour workshift, the products of the measured concentrations and sampling times in minutes are totaled and divided by 480 min

$$\text{TWA} = \frac{(C_1 \times T_1) + (C_2 \times T_2) + \ldots + (C_n \times T_n)}{480 \text{ min}}$$

where C is the concentration in milligrams per cubic meter or parts per million and T is the sampling time in minutes. Results are expressed as an 8-h TWA.

Sometimes sampling cannot be completed for the entire workshift for various reasons even though the worker works the entire shift. Three methods are available for calculating the TWA exposures based on different assumptions:

- **Assume the same exposure for the unsampled period** — If during the unsampled period the worker is performing the same tasks as during the sampled period, the industrial hygienist may assume the same exposures as for the sampled period. To calculate the 8-h TWA, the previous equation is used except now the denominator is the total time sampled rather than 480 min. Results are expressed as an 8-h TWA.

$$\text{TWA} = \frac{(C_1 \times T_1) + (C_2 \times T_2) + \ldots + (C_n \times T_n)}{T_1 + T_2 + \ldots + T_n}$$

- **Assume zero exposure for the unsampled period** — If during the unsampled period the worker is not performing the same task as during the sampled period, the industrial hygienist may assume that the unsampled period has zero exposure. To calculate the 8-h TWA, the first equation is used with a concentration of zero for the unsampled period. Results are expressed as an 8-h TWA.

$$\text{TWA} = \frac{(C_1 \times T_1) + (C_2 \times T_2) + \ldots + (C_n \times T_n) + (0 \times T_{unsampled})}{480 \text{ min}}$$

- **There is not enough information to make either assumption** — If the industrial hygienist is unable to determine whether the exposure during the unsampled period was either zero or the same as the sampled period, then an 8-h TWA cannot be calculated. The time-weighted average exposure can be calculated using the same equation as used for the assumption that the exposure in the unsampled period is the same as in the sampled period. Results are expressed as a TWA and not as an 8-h TWA.

$$\text{TWA} = \frac{(C_1 \times T_1) + (C_2 \times T_2) + \ldots + (C_n \times T_n)}{T_1 + T_2 + \ldots + T_n}$$

Extended shifts longer than 8 h may also be encountered due to overtime work or schedules that require longer shifts. Three methods are available to compare extended shift TWA exposures to PELs and TLVs.

- **Calculate an 8-h TWA** — Using all the data collected for the entire workshift, calculate the 8-h TWA by totalling the products of the measured concentrations and sampling times in minutes for the entire workshift and dividing by 480 min. Results are expressed as an 8-h TWA even though the total sampling time is greater than 8 h. A direct comparison can then be made with the PELs and TLVs.
- **Adjust the standard** — Brief and Scala developed a method of adjusting the PEL or TLV (referred to jointly as the occupational exposure limit [OEL]) to encompass the length of the extended shift. First a TWA exposure is calculated for the entire shift[4].

$$\text{TWA} = \frac{(C_1 \times T_1) + (C_2 \times T_2) + \ldots + (C_n \times T_n)}{T_1 + T_2 + \ldots + T_n}$$

  - Results are expressed as an H-h TWA where H is the number of hours in an extended shift. Next a reduction factor (RF) is calculated to adjust the OEL for the extended shift.

$$RF = \frac{(8/H) \times (24 - H)}{16}$$

  - The reduction factor is then multiplied by the 8-h TWA OEL (OEL_8) to determine the OEL for the extended shift (OEL_H).

$$OEL_H = RF \times OEL_8$$

  - The H-h TWA can then be directly compared to the OEL_H.

[4] Brief, R.S. and Scala, R.A., Occupational exposure limits for novel work schedules, *AIHA Journal*, 36, 467–469, 1975

- **Sample 8 h only** — According to the Occupational Safety and Health Administration (OSHA) policy, the third method is to sample only the 8 h with highest exposure potential.[5] An 8-h TWA can then be calculated as for a standard 8-h workshift, allowing for direct comparison to the PELs and TLVs.

INDUSTRIAL HYGIENE STATISTICAL ANALYSIS

We have examined sampling strategies for obtaining measurements to determine employee exposure to contaminants in the workplace and the different methods for calculating TWAs. It is at this point that the industrial hygienist must use statistics. Whenever physical properties such as noise or airborne contaminants are measured, random error will typically and unavoidably enter into the system. There are two basic areas of statistics that are used to analyze data: descriptive statistics and inferential statistics.

Descriptive statistics are tools used for summarizing data usually with measures of central tendency and dispersion. Measures of central tendency provide characteristic descriptive data such as the mean (average), the median (the value that divides the range of values in half with equal observations above and below this value), and the mode (the value that occurs with the greatest frequency). Also included in descriptive statistics would be measures of dispersion such as the range (the minimum and maximum values for a given sample), the standard deviation (the distribution or dispersion of values around the mean value), and the coefficient of variation (the standard deviation divided by the mean or by a set value, sometimes expressed as a percentage).

A simple example demonstrating central tendency and dispersion could be accomplished by measuring the height of each student in the class. The average height of the students is determined by adding these measured heights, and dividing the total height by the total number of students to yield the average height of students in the class. By calculating the standard deviation, it is possible to determine the distribution of student heights above and below the mean height. By adding and subtracting one standard deviation from the mean, approximately 68% of the student measurements will be represented. By adding and subtracting two standard deviations from the mean value, approximately 95% of the class will be represented. And, finally, by adding and subtracting three standard deviations from the mean, 99% of the class will be represented. By calculating central tendencies and dispersions, industrial hygienists can determine the mean and variability of the measurements of interest, which are indicators of the precision of the techniques used.

Inferential statistics are tools used to assist in drawing conclusions about a population based upon samples taken. As already pointed out in the section on sampling strategies, it is often impractical to obtain continuous samples for every employee in all areas. As a consequence, samples are taken to represent the population of interest. Through inferential statistics, it is possible to establish sound conclusions based upon the samples taken. For example, assume that an industrial hygienist is interested in determining the risk of noise exposure for punch press operators vs. mechanics. Rather than taking full-period single samples of all of the press operators and the mechanics, representatives with maximum risk for both job classifications could be randomly sampled and T-tests performed to determine significant differences between the mean results of both groups. Results from this inferential statistic could assist the industrial hygienist in determining which population to address regarding hearing conservation activities. While perhaps oversimplified, inferential statistics have a significant role to play in industrial hygiene data analysis.

Industrial hygienists use descriptive statistical tools, including central tendency and measurements of dispersion (preferring the terms upper and lower confidence limits to standard deviations), when addressing the issue of variability of sampling measurements. Statistical analysis is extremely important when examining variability, typically referred to as errors, in industrial hygiene measurement and analysis. Errors are the difference between measured values and true values. Relative error is error divided by the true value (expressed as percentages). There are two broad classes of errors: systematic error and random error.

[5] Occupational Safety and Health Administration, Standard Interpretations, http://www.osha-slc.gov/OshDoc/Interp_data/I19991110A.html.

Systematic errors result from the occurrence of known events or laws. They affect the accuracy, but not the precision of the monitoring technique. Examples of systematic errors include improper calibration of an instrument, increases in concentration of a contaminant because of engineering control failure, operator errors or mistakes, and errors of method. Random errors are the result of uncontrollable or unknown variables such as random fluctuation in pump flow rate, sensor accuracy resulting from a bad alkaline battery, variability in airborne concentration because of the production process, or analytical errors resulting from variable laboratory procedures. The total error associated with a sampling result is based upon all the sources of possible errors. This value is determined by taking the square root of the sum of the individual squared errors.

It is important to review descriptive statistics and sampling errors because they form the basis for the concepts associated with SAEs. Industrial hygiene OSHA compliance officers use SAEs to determine whether employee exposures are in or out of compliance. This statistical procedure is presented in the chapter titled "Evaluation of Exposure Levels for Air Contaminants" in the *OSHA Technical Manual*. The following is a summary of SAEs and confidence limits as presented in the OSHA manual for full-period single and consecutive sampling.

- **Definition of SAEs —** When an employee is sampled and the results are analyzed, the measured exposure will rarely be the same as the true exposure. This variation is due to SAEs. The total error is dependent upon the combined effects of the contributing errors inherent in sampling, analysis, and pump flow.
- **Definition of confidence limits —** Error factors determined by statistical methods shall be incorporated into the sample results to obtain the lowest and highest value that the true exposure could be (with a given degree of confidence). The lower value is termed the LCL and the upper value is termed the UCL. These confidence limits are termed one-sided since the only concern is with being confident that the true exposure is on one side of the PEL.
- **Determining SAEs —** SAEs that provide a 95% confidence limit have been developed and listed in chemical information manuals. If there is no SAE listed for a specific substance, apply the manufacturer's recommended error.
- **Confidence limits —** One-sided confidence limits can be used to classify the measured exposure into one of three categories:
  - If the measured results do not exceed the standard and the UCL does not exceed the standard, there is 95% confidence that the company is in compliance (UCL $\leq$ 1 [no violation]).
  - If the measured exposure exceeds the PEL and the LCL of that exposure also exceeds the PEL, there is 95% confidence that the company is in violation (LCL > 1).
  - If the measured exposure does not exceed the PEL, but the UCL does exceed the PEL and/or if the measured exposure exceeds the PEL, but the LCL is below the PEL, the company may not be in compliance (LCL $\leq$ 1 and UCL > 1 [possible overexposure]).
- **Sampling methods —** The LCL and UCL are calculated differently depending upon the type of sampling method used. Sampling methods can be classified into one of three categories.
  - Full-period, continuous single sampling — Full-period, continuous single sampling is defined as sampling over the entire sample period with only one sample. The sampling may be for a full-shift or for a short period ceiling determination.
  - Full-period, consecutive sampling — Full-period, consecutive sampling is defined as sampling using multiple consecutive samples of equal or unequal time duration which, if combined, equal the total duration of the sampling period.
  - Grab sampling — Grab sampling is defined as collecting a number of short-term samples at various times during the sample period which, when combined, provide an estimate of exposure over the total period.
- **Adjustments —** Generally, sampling and calibration are conducted at approximately the same temperature and pressure. Where sampling and calibration are conducted at substantially different temperatures and pressures, an adjustment to the measured air volume may be required.
- **Calculation method for full-period single sampling —** Use the following method to calculate full-period single sampling:
  - Obtain the full-period sampling result (value X), the PEL, and the SAE.

- Divide X by the PEL to determine Y, the standardized concentration.

$$Y = \frac{X}{PEL}$$

- Compute the UCL (95%).

$$\text{UCL (95\%)} = \text{Y} + \text{SAE}$$

- Compute the LCL (95%).

$$\text{LCL (95\%)} = \text{Y} - \text{SAE}$$

- Classify the exposure according to the following classification system:
  - UCL ≤ 1 = no violation
  - LCL ≤ 1 and UCL > 1 = possible overexposure
  - LCL > 1 = violation exists
- **Calculation method for full-period consecutive sampling** — The use of multiple consecutive samples will result in slightly lower SAEs than the use of one continuous sample since the inherent errors tend to partially conceal each other. The mathematical calculations are more complex:
  - Obtain the sample values (X_1, X_2, X_3, etc.), their time durations (T_1, T_2, T_3, etc.), and the SAE.
  - Compute the TWA exposure.

$$\text{TWA} = \frac{(X_1 \times T_1) + (X_2 \times T_2) + \ldots + (X_n \times T_n)}{8\ h}$$

- Divide the TWA exposure by the PEL to find Y.

$$Y = \frac{TWA}{PEL}$$

- Compute the UCL (95%).

$$\text{UCL (95\%)} = \text{Y} + \text{SAE}$$

- Compute the LCL (95%).

$$\text{LCL (95\%)} = \text{Y} - \text{SAE}$$

- Classify the exposure according to the following classification system:
  - UCL ≤ 1 = no violation
  - LCL ≤ 1 and UCL > 1 = possible overexposure
  - LCL > 1 = violation exists

- When the LCL ≤ 1.0 and UCL > 1.0, the results are in the possible overexposure region and the data must be analyzed using the more exact calculation for the full-period consecutive sampling.

$$LCL = Y - SAE\frac{\sqrt{\left[(T_1)^2 \times (X_1)^2\right] + \cdots + \left[(T_n)^2 \times (X_n)^2\right]}}{PEL \times (T_1 + \cdots + T_n)}$$

Statistics are also necessary when performing a quantitative exposure assessment to determine the acceptability of the exposures. Some of these statistics can be calculated by hand or with a calculator, but some may also require the use of a computer. A spreadsheet with the necessary calculations preprogrammed is included in the book by Mulhausen and Damiano (see Bibliography). Other programs are also available that can perform the necessary statistical calculations.

During the early stages of comprehensive exposure assessment, workers are divided into similar exposure groups (SEGs) based on information collected such as job title, job task, workshift, and stressor. The goal of a comprehensive exposure assessment is to achieve acceptable exposures for each stressor in each SEG. Workplace monitoring is conducted, as necessary, for the stressors within an SEG. Workers in an SEG are randomly selected and monitoring is performed randomly throughout a specified time period, such as one year. A minimum of six exposures (8-h TWAs, 10-h TWAs, 15-min samples, etc.) with the same averaging time is recommended before performing the statistics. Some type of OEL, whether PEL, TLV, or in-house standard, is also necessary for the acceptability determination.

The first step is to determine whether the collected exposures are lognormally distributed, as is common with industrial hygiene data. The Shapiro-Wilk's W-test is one method for making the determination. As long as the exposures are lognormally distributed, the industrial hygienist then has multiple statistics to use in determining acceptability. In order to be considered, acceptable exposures must be less than the appropriate OEL. Upper confidence limits (UCLs) are used for the determination to help account for variability and uncertainty. Each company or organization needs to decide at what levels UCLs are low enough to be considered acceptable as well as at what level they are unacceptable. Acceptability could be defined as exposures with upper tolerance limits (UTLs) less than the OEL. The UTL is generally the most extreme value. Other possiblilties could include the exceedance fraction must be less than 5% (i.e., the probability of an exposure exceeding the OEL is less thean 5%); the arithmetic mean of all exposures, the geometric mean of all exposures, and the minimum variance unbiased estimate (MVUE) and its Land's Exact UCL must all be below the OEL. Unacceptable exposures may be defined as any time an individual or average exposure exceeds the OEL. Uncertin exposures then are those that do not meet either the acceptable definition or the unacceptable definition.

APPENDIX A — EXAMPLES

Example A.1

You have completed monitoring of a work area for respirable dust by taking two samples of 3 h and one sample of 2 h. After post-weighing your filters, you determine the following concentrations for each sample period:

Concentration	Time Samples
2.58 mg/m^3	3 h
1.51 mg/m^3	2 h
3.53 mg/m^3	3 h

Given an SAE of 0.21, determine whether this employee's exposure was in or out of compliance for the OSHA-PEL for respirable dust (PEL = 5 mg/m³).

Solution A.1

$$8\text{-}h\ TWA = \frac{(C_1T_1)+(C_2T_2)+(C_3T_3)}{8\ h}$$

$$= \frac{(2.58\times 3)+(1.51\times 2)+(3.53\times 3)}{8\ h}$$

$$= 2.67\ \text{mg / m}^3$$

$$Y = \frac{TWA}{PEL}$$

$$= \frac{2.67}{5.0}$$

$$= 0.53$$

$$UCL_{95\%} = Y + SAE$$

$$= 0.53 + 0.21$$

$$= 0.74$$

$$LCL_{95\%} = Y - SAE$$

$$= 0.53 - 0.21$$

$$= 0.32$$

Since $UCL_{95\%}$ is less than 1, the employee's exposure is in compliance.

Example A.2

You have been monitoring a compressed air spray painting operation for toluene. To date you have ten 15-min average concentrations. The short-term exposure limit (STEL) for toluene is 560 mg/m³. The measured concentrations are as follows (in milligrams per cubic meter). Would this operation be considered acceptable?

8.5, 2.0, 3.5, 4.0, 4.0, 10.7, 6.9, 3.3, 3.3, 3.3

Solution A.2

The data were entered into the spreadsheet from the Mulhausen and Damiano book (see printout below). First you verify that the data are lognormally distributed. Under the heading "Test for Distribution Fit," the answer next to lognormal is yes. The lognormal parametric statistics can be utilized in the determination of acceptability.

Next you begin comparing certain statistics to the STEL of 560 mg/m³:

- The mean is 4.95 mg/m^3.
- The geometric mean is 4.368 mg/m^3.
- The MVUE is 4.910 mg/m^3.
- The Land's Exact UCL is 7.260 mg/m^3.
- The UTL is 19.426 mg/m^3.

All of these values are less than the STEL, by at least a factor of 28. In addition, the exceedance fraction (percent above OEL under the heading "Lognormal Parametric Statistics") is 0, which is less than 5%. All of the statistics agree that the short-term exposures to toluene during compressed air spray painting are acceptable.

Industrial Hygiene Statistics

Data Description:

OEL
560

Sample Data (max n = 50) No less-than (<) or greater-than (>)
8.5
2
3.5
4
4
10.7
6.9
3.3
3.3
3.3

Descriptive Statistics	
Number of samples (n)	10
Maximum (max)	10.7
Minimum (min)	2
Range	8.7
Percent above OEL (%>OEL)	0.000
Mean	4.950
Median	3.750
Standard deviation (s)	2.794
Mean of logtransformed data (LN)	1.474
Std. deviation of logtransformed data (LN)	0.513
Geometric mean (GM)	4.368
Geometric standard deviation (GSD)	1.670
Test For Distribution Fit	
W-test of logtransformed data (LN)	0.902
Lognormal (a = 0.05)?	Yes
W-test of data	0.818
Normal (a = 0.05)?	No
Lognormal Parametric Statistics	
Estimated Arithmetic Mean - MVUE	4.910
$LCL_{1,95\%}$ - Land's "Exact"	3.781
$UCL_{1,95\%}$ - Land's "Exact"	7.260
95th Percentile	10.151
$UTL_{95\%,95\%}$	19.426
Percent above OEL (%>OEL)	0.000
$LCL_{1,95\%\%}$>OEL	<0.1
$UCL_{1,95\%}$%>OEL	<0.000
Normal Parametric Statistics	
Mean	4.950
$LCL_{1,95\%}$ - t statistics	3.331
$UCL_{1,95\%}$ - t statistics	6.569
95th Percentile - Z	9.546
$UTL_{95\%,95\%}$	13.08
Percent above OEL (%>OEL)	0.000

APPENDIX A — CASE STUDY

Case Study

A dry cleaning operation has been monitored for perchloroethylene. One person is responsible for operating four generation-3 dry-to-dry machines. Generation-3 machines are equipped with refrigerated condensers that recirculate the heated drying air through a vapor condenser, but are not equipped with local exhaust ventilation. The machines have either a 40-lb capacity (two machines) or a 70-lb capacity (two machines).

The worker has been monitored 6 times in the past 6 months with the following results (in milligrams per cubic meter 8-h TWAs).

78, 43, 145, 110, 91, 59

The company uses the 1989 PEL for perchloroethylene (170 mg/m³ 8-h TWA) as their in-house standard. The company has an established comprehensive exposure assessment strategy (acceptable = UTL < AL [action level] and F [exceedance fraction] < 5%; unacceptable = UTL > OEL) and also performs compliance calculations on the highest exposures (SAE = 0.09).

The highest exposure for this 6-month period was 145 mg/m³.

$$Y = \frac{145}{170}$$

$$= 0.85$$

$$UCL_{95\%} = 0.85 + 0.09$$

$$= 0.94$$

Since $UCL_{95\%}$ is less than 1, the highest exposure is in compliance.

Looking at all six exposures using the spreadsheet from the Mulhausen and Damiano book, we find that the data are lognormally distributed. The exceedance fraction is 4.5%, which meets the acceptable requirements, but the UTL is 408 mg/m³, which is unacceptable. The mean (87.7 mg/m³), geometric mean (81.2 mg/m³), MVUE (87.8 mg/m³), Land's Exact UCL (144 mg/m³), and the 95th percentile (166 mg/m³) all are either less than the action level (85 mg/m³) or between the action level and OEL. Therefore, the exposure assessment for the dry cleaning operation using perchloroethylene could be listed as uncertain, requiring additional information.

Industrial Hygiene Statistics

Data Description:

OEL
170

Sample Data (max n = 50) No less-than (<) or greater-than (>)
78
43
145
110
91
59

Descriptive Statistics	
Number of samples (n)	6
Maximum (max)	145
Minimum (min)	43
Range	102
Percent above OEL (%>OEL)	0.000
Mean	87.667
Median	84.500
Standard deviation (s)	36.626
Mean of logtransformed data (LN)	4.397
Std. deviation of logtransformed data (LN)	0.436
Geometric mean (GM)	81.227
Geometric standard deviation (GSD)	1.546

Test For Distribution Fit	
W-test of logtransformed data (LN)	1.003
Lognormal (a = 0.05)?	Yes
W-test of data	0.990
Normal (a = 0.05)?	Yes

Lognormal Parametric Statistics	
Estimated Arithmetic Mean - MVUE	87.836
$LCL_{1,95\%}$ - Land's "Exact"	65.316
$UCL_{1,95\%}$ - Land's "Exact"	144.423
95th Percentile	166.325
$UTL_{95\%,95\%}$	408.426
Percent above OEL (%>OEL)	4.502
$LCL_{1,95\%\%}$>OEL	0.309
$UCL_{1,95\%}$%>OEL	28.842

Normal Parametric Statistics	
Mean	87.667
$LCL_{1,95\%}$ - t statistics	57.537
$UCL_{1,95\%}$ - t statistics	117.797
95th Percentile - Z	147.916
$UTL_{95\%,95\%}$	223.44
Percent above OEL (%>OEL)	1.229

APPENDIX A — PROBLEM SET

1. Three instructors at a firing range were monitored for lead. Instructor A worked an 8-h shift and was monitored the entire day. Instructor B had a doctor's appointment and was monitored for the 5 h he was present. Instructor C tried to avoid the industrial hygenist and entered the firing range before being assigned a sampling pump. Instructor C was monitored for 6 of the 8 h he was present.

a. Calculate the 8-h TWA for Instructor A given the following measured concentrations:

Concentration	Time Sampled
0.025 mg/m^3	3 h
0.043 mg/m^3	2 h
0.019mg/m^3	3 h

b. Calculate the 8-h TWA for Instructor B given the following measured concentrations:

Concentration	Time Sampled
0.031 mg/m^3	3 h
0.024 mg/m^3	2 h

c. Calculate the 8-h TWA for Instructor C given the following measured concentrations:

Concentration	Time Sampled
0.042 mg/m^3	3 h
0.047 mg/m^3	3 h

2. Given the SAE for lead is 0.11 and the PEL is 0.05 mg/m^3 8-h TWA.
   a. Determine whether Instructor A is in or out of compliance.
   b. Determine whether Instructor B is in or out of compliance.
   c. Determine whether Instructor C is in or out of compliance.
3. On a separate day another instructor, Instructor D, was tasked with cleaning the bullet trap. The operation took 10 h and the entire operation was sampled with the following results.

Concentration	Time Sampled
0.06 mg/m^3	105 min
0.08 mg/m^3	113 min
0.03 mg/m^3	132 min
0.04 mg/m^3	130 min
0.01 mg/m^3	120 min

a. Calculate the 8-h TWA using all of the data. Do the results exceed the PEL?
b. Calculate the 10-h TWA. Does the result exceed the Brief and Scala adjusted OEL?
c. Calculate the 8-h TWA using the OSHA approach. Does the result exceed the PEL?
d. Can OSHA write a citation for noncompliance?

APPENDIX B

Example: Outlined Format of an Industrial Hygiene Evaluation Report

This represents one example of an outlined format of sections and information that should be summarized and documented in an industrial hygiene evaluation report. The format and content of reports may vary depending on the intended audience and the type of sampling and analysis conducted.

- Overview/Executive Summary
  - Where, when, and by whom was sampling conducted?
  - What was sampled and analyzed?
  - Which personnel and areas were sampled?
  - Summary of operations/processes
  - Summary of results relative to applicable Occupational Safety and Health Administration permissible exposure limits (OSHA-PELs), American Conference of Governmental Industrial Hygienists threshold limit values (ACGIH-TLVs), and other applicable occupational exposure limits (e.g., National Institute for Occupational Safety and Health recommended exposure limits [NIOSH-RELs]; Mine Safety and Health Administration threshold limit values [MSHA-TLVs])
- Materials and Methods
  - Instruments and media used for calibration, sampling, and analysis
  - Calibrated flow rates for both pre- and post-sampling
  - Cited methods (e.g., NIOSH, OSHA, other)
  - Identification of laboratory
  - Duration of sampling (e.g., full-shift, partial shift, extended shift)
- Results
  - Tables summarizing calibration, sampling, and analytical data
  - Succinct, factual statements based on observations and measurements
- Discussion and Conclusions
  - Significance of findings
- Recommendations
  - Control measures (e.g., work practices, ventilation, personal protection equipment [PPE] etc.)
  - Follow-up sampling and analysis
  - Record keeping
- Appendices
  - Field monitoring forms
  - Other applicable information (as needed)

APPENDIX C

Answers to Case Studies for Units 4, 5, 6, 7, and 9 and Problem Sets in Appendix A

ANSWERS TO UNIT 4 CASE STUDY (DATA AND CALCULATION TABLES ONLY)

Table 4.1 Calibration and Field Data

Name of Personnel or Area	Sample No.	Analyte	Avg. Pre-sample Q(l/min)	Avg. Post-sample Q(l/min)	Avg. Q (l/min)	Sample Start Time	Sample Stop Time	Sample Time (min)	Air Volume Sampled (l)
Jim K.	2155	Total dust	Pump No. 3021: 2.00	Pump No. 3021: 2.00	**2.00**	0700	1059	**239**	**478**
	2157					1100	1500	**240**	**480**
Mixer 1	2136	Total dust	Pump No. 3023: 2.01	Pump No. 3023: 2.04	**2.03**	0701	1101	**240**	**487.2**
	2135					1103	1459	**236**	**479.1**
Blank	2130	Total dust	N/A	N/A	**N/A**	N/A	N/A	**0**	**0**

N/A = Not applicable.

Table 4.2 Lab Data

Name of Personnel or Area	Sample No.	Analyte	Pre-sample Weight (mg)	Post-sample Weight (mg)	Weight Total Dust (mg)	Percent Crystalline Quartz (%)	AirVolume Sampled (m^3)	Conc. Total Dust (mg/m^3)
Jim K.	2155	Total dust	11.016	15.643	**4.627**	N/A	**0.478**	**9.68**
	2157		11.000	15.444	**4.444**	N/A	**0.480**	**9.26**
Mixer 1	2136	Total dust	11.101	13.756	**2.655**	N/A	**0.4872**	**5.45**
	2135		11.099	14.041	**2.942**	N/A	**0.4791**	**6.14**
Blank	2130	Total dust	11.011	11.011	<LOD	N/A	**0**	**<LOD**

LOD = Limit of detection.
N/A = Not applicable.

Table 4.3 Calculated Time-Weighted Averages

Name of Personnel or Area	Sample No.	Analyte	8-hr TWA Total Dust (mg/m³)
Jim K.	2155	Total dust	**9.45**
	2157		
Mixer 1	2136	Total dust	**5.74**
	2135		
Blank	2130	Total dust	**<LOD**

LOD = Limit of detection.

ANSWERS TO UNIT 5 CASE STUDY (DATA AND CALCULATION TABLES ONLY)

Table 5.2 Calibration and Field Data

Name of Personnel or Area	Sample No.	Analyte	Avg. Pre-sample Q (l/min)	Avg. Post-sample Q (l/min)	Avg. Q (l/min)	Sample Start Time	Sample Stop Time	Sample Time (min)	Air Volume Sampled (l)
Tim D.	0321	Respirable dust and quartz	Pump No. 221: 1.70	Pump No. 221: 1.72	**1.71**	0700	1100	**240**	**410.4**
	0323					1101	1500	**239**	**408.7**
Mike K.	0329	Respirable dust and quartz	Pump No. 225: 1.69	Pump No. 225: 1.72	**1.71**	0701	1102	**241**	**412.1**
	0333					1103	1501	**238**	**407.0**
Blank	0300	Respirable dust and quartz	N/A	N/A	**N/A**	N/A	N/A	**0**	**0**

N/A = Not applicable.

Table 5.3 Lab Data

Name of Personnel or Area	Sample No.	Analytes	Pre-sample Weight (mg)	Post-sample Weight (mg)	Weight Respirable Dust (mg)	Percent Crystalline Quartz (%)	Air Volume Sampled (m³)	Conc. Respirable Dust (mg/m³)	Conc. Respirable Quartz Dust (mg/m³)
Tim D.	0321	Respirable dust and quartz	12.056	13.964	**1.908**	**8**	**0.4104**	**4.65**	**0.37**
	0323		11.999	14.004	**2.005**	**8**	**0.4087**	**4.91**	**0.39**
Mike K.	0329	Respirable dust and quartz	12.115	13.075	**0.960**	**8**	**0.4121**	**2.33**	**0.19**
	0333		12.104	13.604	**1.500**	**8**	**0.407**	**3.69**	**0.30**
Blank	0300	Respirable dust and quartz	12.223	12.223	**<LOD**	**<LOD**	**0**	**<LOD**	**<LOD**

LOD = Limit of detection.

N/A = Not applicable.

Table 5.4 Calculated Time-Weighted Averages

Name of Personnel or Area	Sample No.	Analytes	8-h TWA Respirable Dust (mg/m³)	8-h TWA Respirable Quartz Dust (mg/m³)
Tim D.	0321 0323	Respirable dust and quartz	**4.77**	**0.38**
Mike K.	0329 0333	Respirable dust and quartz	**3.00**	**0.24**
Blank	s0300	Respirable dust and quartz	**<LOD**	**<LOD**

ANSWERS TO UNIT 6 CASE STUDY (DATA AND CALCULATION TABLES ONLY)

Table 6.1 Calibration and Field Data — Asbestos Fibers

Name of Personnel or Area	Sample No.	Analyte	Avg. Pre-sample Q (l/min)	Avg. Post-sample Q (l/min)	Avg. Q (l/min)	Sample Start Time	Sample Stop Time	Sample Time (min)	Air Volume Sampled (l)
Gary M.	0052	Fibers	Pump No. 133: 2.25	Pump No. 133: 2.25	**2.25**	1503	1930	**267**	**600.8**
	0055					1931	2300	**209**	**470.3**

Table 6.2 Lab Data — Asbestos Fibers

Name of Personnel or Area	Sample No.	Analyte	Fibers Counted Per Total Fields (f/field)	Fiber Density (f/mm²)	Air Volume Sampled (cm³)	Conc. Fibers (f/cm³)
Gary M.	0052	Fibers	100/67	**193.89**	**$6.008 \bullet 10^5$**	**0.12**
	0055		82/100	**106.53**	**$4.703 \bullet 10^5$**	**0.09**

Lab Notes: (1) Calibration of the Walton-Beckett graticule using a stage micrometer indicated that the measured diameter was 99.0 µm. (2) Assume total exposed surface of 25 mm diameter MCE filter during sampling was 385 mm². (3) Assume all field blanks were <limit of detection (LOD).

Table 6.3 Calculated Time-Weighted Averages

Name of Personnel or Area	Sample No.	Analyte	8-h TWA Asbestos Fibers (f/cm³)
Gary M.	0052 0055	Fibers	**0.11**
Blank	0500	Fibers	**<LOD**

ANSWERS TO UNIT 7 CASE STUDY (DATA AND CALCULATION TABLES ONLY)

Table 7.1 Calibration and Field Data

Name of Personnel or Area	Sample No.	Analyte	Avg. Pre-sample Q (l/min)	Avg. Post-sample Q (l/min)	Avg. Q (l/min)	Sample Start Time	Sample Stop Time	Sample Time (min)	Air Volume Sampled (l)
Joe M.	1136	Lead	Pump No. 001: 2.16	Pump No. 001: 2.20	**2.18**	1500	1730	**150**	**327.0**
	1140					1730	2000	**150**	**327.0**
	1128					2000	2300	**180**	**392.4**
Tank 1 Railing	1138	Lead	Pump No. 005: 2.08	Pump No. 005: 2.02	**2.05**	1500	1900	**240**	**492.0**
	1125					1900	2300	**240**	**492.0**
Blank	1131	Lead	N/A	N/A	**N/A**	N/A	N/A	**N/A**	—

N/A = Not applicable.

Table 7.2 Lab Standard Curve Data for Atomic Absorption Spectrometer

Abs	0	0.1	0.2	0.3	0.4	0.5	0.6	0.7	0.8	0.9	1.0
Lead Std. (μg/ml)	0	0.25	0.50	0.75	1.00	1.25	1.50	1.75	2.00	2.25	2.50

Table 7.3 Lab Data

Name of Personnel or Area	Sample No.	Analyte	Absorbance	Conc. Analyte (μg/ml)	Volume Acid to Dissolve Filter (ml)	Mass of Analyte (mg)	Air Volume Sampled (m^3)	Conc. Lead Dust (mg/m^3)
Joe M.	1136	Lead	0.70	**1.75**	10	**0.018**	**0.327**	**0.06**
	1140		0.50	**1.25**		**0.013**	**0.327**	**0.04**
	1128		0.80	**2.00**		**0.020**	**0.3924**	**0.05**
Tank 1 Railing	1138	Lead	0.40	**1.00**	10	**0.010**	**0.492**	**0.02**
	1125		0.30	**0.75**		**0.008**	**0.492**	**0.02**
Blank	1131	Lead	<LOD	**N/A**	10	**<LOD**	**0**	**<LOD**

LOD = Limit of detection.
N/A = Not applicable.

Table 7.4 Calculated Time-Weighted Averages

Name of Personnel or Area	Sample No.	Analyte	8-h TWA Lead Dust (mg/m^3)
Joe M.	1136	Lead Dust	**0.05**
	1140		
	1128		
Tank 1 Railing	1138	Lead Dust	**0.02**
	1125		
Blank	1131	Lead Dust	**<LOD**

ANSWERS TO UNIT 9 CASE STUDY (DATA AND CALCULATION TABLES ONLY)

Table 9.1 Calibration and Field Data

Name of Personnel or Area	Sample No.	Analytes	Avg. Pre-sample Q (cm³/min)	Avg. Post-sample Q (cm³/min)	Avg. Q (cm³/min)	Sample Start Time	Sample Stop Time	Sample Duration (min)	Air Volume Sampled (l)
	132		Pump No. 342	Pump No. 342		1500	1700	**120**	**6.12**
	126					1700	1900	**120**	**6.12**
Tim P.	136	Toluene	50	52	51	1900	2100	**120**	**6.12**
	128					2100	2300	**120**	**6.12**
Blank	105	Toluene	NA	NA	N/A	N/A	N/A	**N/A**	**0**

N/A = Not applicable.

Table 9.2 Lab Data

Name of Personnel or Area	Sample No.	Analytes	Mass Analyte Detected Front (mg)	Mass Analyte Detected Back (mg)	Mass Analyte Detected Total (mg)	Air Volume Sampled (m³)	Conc. Analyte (mg/m³)	Conc. Analyte (ppm)
	132		1.9	0.1	**2.0**	**0.0061**	**327.87**	**87.14**
	126		2.5	<LOD	**2.5**	**0.0061**	**409.84**	**108.92**
Tim P.	136	Toluene	2.9	<LOD	**2.9**	**0.0061**	**475.41**	**126.35**
	128		3.3	<LOD	**3.3**	**0.0061**	**540.98**	**143.77**
Blank	105	Toluene	<LOD	<LOD	**<LOD**	**N/A**	**<LOD**	**<LOD**

LOD = Limit of detection.

N/A = Not applicable.

Table 9.3 Calculated Time-Weighted Averages

Name of Personnel or Area	Sample No.	Analytes	8-h TWA (ppm)
	132		
	126		
T. Pender	136	Toluene	**116.55**
	128		
Blank	105	Toluene	**<LOD**

ANSWERS TO APPENDIX A PROBLEM SET

1a.

$$8\text{-h TWA} = \frac{(3\times 0.025)+(2\times 0.043)+(3\times 0.019)}{8}$$

$$= 0.027 \text{ mg/m}^3$$

1b.

Assume zero for unsampled period:

$$8\text{-h TWA} = \frac{(3\times 0.031)+(2\times 0.024)+(3\times 0)}{8}$$

$$= 0.018 \text{ mg/m}^3$$

1c.

Assume same exposure for unsampled period:

$$8\text{-h TWA} = \frac{(3\times 0.042)+(3\times 0.047)}{6}$$

$$= 0.044 \text{ mg/m}^3$$

2a.

$$UCL_{95} = \frac{0.027}{0.05} + 0.11$$

$$= 0.65 \ (< 1, \textit{ in compliance})$$

2b.

$$UCL_{95} = \frac{0.018}{0.05} + 0.11$$

$$= 0.47 \ (< 1, \textit{ in compliance})$$

2c.

$$UCL_{95} = \frac{0.044}{0.05} + 0.11$$

$$= 0.99 \ (< 1, \textit{ in compliance})$$

3a.

$$8\text{-h TWA} = \frac{(0.06\times105)+(0.08\times113)+(0.03\times132)+(0.04\times130)+(0.01\times120)}{480}$$

$$= 0.053 \text{ mg/m}^3 \text{(exceeds PEL)}$$

3b.

$$10\text{-h TWA} = \frac{(0.06\times105)+(0.08\times113)+(0.03\times132)+(0.04\times130)+(0.01\times120)}{600}$$

$$= 0.043 \text{ mg/m}^3 \text{ (exceeds PEL}_{10}\text{)}$$

$$\text{RF} = \frac{(8/10)\times(24-10)}{16}$$

$$= 0.7$$

$$\text{PEL}_{10} = 0.7\times0.05$$

$$= 0.035 \text{ mg/m}^3$$

3c.

$$8\text{-h TWA} = \frac{(0.06\times105)+(0.08\times113)+(0.03\times132)+(0.04\times130)}{480}$$

$$= 0.051 \text{ mg/m}^3$$

3d.

$$LCL_{95} = \frac{0.051}{0.05} - 0.11$$

$$= 0.91 \text{ (possible overexposure, cannot cite)}$$

Bibliography and Recommended References

Abramowitz, M., *Contrast Methods in Microscopy: Transmitted Light*, Vol. 2, Olympus Corporation, Lake Success, NY, 1987.

Berger, E.H., Royster, L.H., Royster, J.D., Driscoll, D.P., and Layne, M., Eds., *The Noise Manual*, 5th ed., AIHA Press, Akron, OH, 2000.

Cassinelli, M.E. and O'Connor, P.F., Eds., *NIOSH Manual of Analytical Methods* (NMAM), 4th ed., DHHS and NIOSH Publication No. 94-113, U.S. Government Printing Office, Washington, D.C., 1994.

Chou, J., *Hazardous Gas Monitors — A Practical Guide to Selection, Operation and Applications*, McGraw-Hill Book Company, New York, NY, 2000.

Cohen, B.S. and McCammon, C.S., Eds., *Air Sampling Instruments for Evaluation of Atmospheric Contaminants*, 9th ed., The American Conference of Governmental Industrial Hygienists, Cincinnati, OH, 2001.

DiNardi, S., Ed., *The Occupational Environment: Its Evaluation and Control*, AIHA Press, Fairfax, VA, 1997.

Eastman Kodak Company, The Human Factors Section, Health, Safety and Human Factors Laboratory, *Ergonomic Design for People at Work, Vol. 1, Workplace, Equipment, and Environmental Design and Information Transfer*, Van Nostrand Reinhold, New York, NY, 1989.

Eastman Kodak Company, The Ergonomics Group, Health and Environmental Laboratories, *Ergonomic Design for People at Work, Vol. 2, The Design of Jobs, including Work Patterns, Hours of Work, Manual of Materials Handling Tasks, Methods to Evaluate Job Demands, and the Physiological Basis of Work*, Van Nostrand Reinhold, New York, NY, 1989.

Gollnick, D.A., *Basic Radiation Protection Technology*, 4th ed., Pacific Radiation Corporation, Altadena, CA, 2000.

Kenkel, J., *Analytical Chemistry Refresher Manual*, Lewis Publishers, Inc., Chelsea, MI, 1992.

Macher, J.M., Ammann, H.A., Burge, H.A., Milton, D.K., and Morey, P.R., Eds., *Bioaerosols: Assessment and Control*, The American Conference of Governmental Industrial Hygienists, Cincinnati, OH, 1999.

Malansky, C.J. and Malansky, S.P., *Air Monitoring Instrumentation: A Manual for Emergency, Investigatory, and Remedial Responders*, Van Nostrand Reinhold, New York, NY, 1993.

Mulhausen J.R. and Damiano J., *A Strategy for Assessing and Monitoring Occupational Exposures*, AIHA Press, Fairfax, VA, 1998.

Plog, B.A., Ed., *Fundamentals of Industrial Hygiene*, 5th ed., National Safety Council, Chicago, IL, 2002.

The American Conference of Governmental Industrial Hygienists, *Threshold Limit Values and Biological Exposure Indices for 2002*, ACGIH, Cincinnati, OH, 2002.

The American Conference of Governmental Industrial Hygienists, *Documentation of the Threshold Limit Values and Biological Exposure Indices*, 7th ed., ACGIH Worldwide, Cincinnati, OH, 2001.

The American Conference of Governmental Industrial Hygienists, *Industrial Ventilation: A Manual of Recommended Practice*, 24th ed., ACGIH Worldwide, Cincinnati, OH, 2001.

The American Industrial Hygiene Association, *Fundamentals of Analytical Procedures in Industrial Hygiene*, AIHA, Akron, OH, 1987.

The American Thoracic Society, *Standardization of Spirometry — 1994 Update, Respiratory Care*, 152, 1107–1136, 1995.

The American Thoracic Society, *AARC Clinical Practice Guideline Spirometry — 1996 Update, Respiratory Care*, 41, 629–636, 1996.

The Occupational Safety and Health Administration, *OSHA Technical Manual*, Commerce Clearing House, Chicago, IL, 1999.

The Occupational Safety and Health Administration, Code of Federal Regulations, Title 29 Sections 1910.134 and 1910.1000, Government Institutes, Rockville, MD, 1993.

The U.S. Department of Health, Education, and Welfare/National Institute for Occupational Safety and Health, *Occupational Exposure Sampling Strategy Manual*, DHEW/NIOSH Publication No. 77-173, U.S. Government Printing Office, Washington, D.C., 1977.

Vincent, J.H., Ed., *Particle Size Selective — Sampling for Particulate Air Contaminants*, ACGIH, Cincinnati, OH, 1999.

Wight, G.D., *Fundamentals of Air Sampling*, Lewis Publishers, Inc., Chelsea, MI, 1994.

INDEX

H

I

L

M

N

O

P

Q

R

S

T

V

W